D0778638

E 40418

Earthquake Protection
of Essential
Building Equipment

Earthquake Protection
of Essential
Building Equipment

Design
Engineering
Installation

Gary L. McGavin

Member, Earthquake Engineering Research Institute
Associate Member, American Institute of Architects

A WILEY-INTERSCIENCE PUBLICATION

JOHN WILEY & SONS   New York   •   Chichester   •   Brisbane   •   Toronto

Library of Congress Cataloging in Publication Data:

McGavin, Gary L.   1948–
    Earthquake protection of essential building equipment.

    "A Wiley-Interscience publication."
    Bibliography: p.
    Includes index.
    1.   Earthquakes and building.   2.   Buildings—Me-
chanical equipment.   I.   Title.

TH1095.M23   696'.028'9   80-23067
ISBN 0-471-06270-7

Printed in the United States of America

10 9 8 7 6 5 4 3 2 1

To Charles W. Roberts and my wife Cathy:
Their counsel and efforts made the task a joy

Preface

Building codes currently in use throughout the United States proclaim that the availability of certain classes of facilities are required for the "good of the public" after major disasters such as earthquakes. These codes specifically require the operability of the facilities without providing a detailed mechanism for obtaining it. Both the structural systems and selected nonstructural components are addressed by the codes. The majority of the equipment items, however, are virtually neglected. The mere written requirement for a facility's operability is not nearly the same as providing a viable means for obtaining it.

This book addresses that perplexing problem: how to maintain operability of equipment after a major earthquake. The program is often more complicated than simple anchorage. Some critical types of equipment, for example, are likely to fail operationally (false signaling of switches, etc.) even if adequate base anchorage has been provided. It is the goal of this book to provide the reader with a plan whereby equipment can be classified and subsequently qualified for the postulated seismic environment in a manner that best suits the individual piece of equipment. In some cases this qualification can be accomplished only by a sophisticated seismic testing program, while at the other end of the spectrum, equipment sometimes may be qualified simply by adequate architectural detailing. All too often, professional design teams and owners rely on an electric plug and gravity to keep a critical piece of equipment in place during an earthquake and functional afterward. Obviously, this is a less than desirable situation.

This book considers equipment from the systems point of view and suggests qualification methods that are suited for each individual piece of equipment and its own particular installation and use. The process is not new to those who have been involved with the design and construction of critical facilities such as nuclear power plants. The detailed approach to equipment qualification is, however, likely to be a fresh idea to those involved with general construction.

The qualification procedures presented herein include:

● Seismic qualification methods.
  • Testing.
  • Analysis.
  • Designer judgment.

- • Prior experience.
- • Combined methods.
- Design earthquakes.
- Seismic categories for equipment.
- Design specification models.
- Suggested diagrammatic installation details.

The detailed discussion of these topics leaves the designer with a complete course of action for qualifying all types of equipment, whether it be critical, support, or miscellaneous equipment.

The handbook format of this book lends itself well to a wide spectrum of users. Design professionals such as architects and electrical, mechanical, plumbing, and structural engineers will find this book useful, as will building owners, facility operators, equipment manufacturers, building departments, inspectors, and students. It has been written for all those interested in seeing any facility remain operational after an earthquake.

The application of the proposals within the book will, according to my colleague Paul Turkheimer, "increase the margin of safety for facilities" that we need to cope with the physical and psychological stresses of a major earthquake. We, the public, cannot tolerate, for instance, a hospital that is unable to accept and treat victims because the emergency power supply system is inoperable as a result of any number of "minor" failures.

The purpose of this book is to provide the path by which the goals of the building codes can be satisfied. Facilities can remain operational after an earthquake if prudent design concepts are employed. The techniques are available, all we need do is utilize them.

GARY L. MCGAVIN

Riverside, California
February 1981

Acknowledgments

There are many individuals, institutions, and businesses that deserve acknowledgment for assistance in the writing of this book. Their valuable time and material have contributed greatly to the research for the text, in some cases as long as 3 years prior to publication. Following is a list of those who have provided this type of support. If it is not complete I apologize to those who were omitted. To all of those who provided support, thank you.

Alyeska Pipeline Service Company

GEORGEANN B. AUSTIN, Stone, Marraccini and Patterson

ROBERT BARNECUT, Stone, Marraccini and Patterson

DOUG BEETHAM, Orange County Department of Architectural Services

DONALD BENKERT, California Dynamics Corporation

DR. RONALD CARLYLE, Cal-Poly Pomona, Department of Civil Engineering

STEVE CARROLL, St. Joseph's Hospital, Santa Ana

PROFESSOR RICHARD CHYLINSKI, Cal-Poly Pomona, Department of Architecture

CHARLES CLAVIN, San Bernardino County

Conwed Corporation

DONALD DOSS, Alcan Aluminum Ceiling Systems

EBCO Manufacturing Company

Earthquake Engineering Research Institute

DICK ENCINAS, Oliver & Williams Elevator Company

CLIFFORD ENDSLEY, Orange County Department of Architectural Services

CHARLES GEORGETTE, Riverside County General Hospital

LT. GRAY, San Bernardino County Sheriff's Department

Grinnell Fire Protection Systems Company, Inc.

Groen Corporation

JAMES F. HALSEY, Brown and Root

PETER L. HANSEN, Architect

x   Acknowledgments

JAMES R. HARRIS, N.B.S. Center for Building Technology

Hayakawa Associates, Los Angeles, California

DONALD HENLEY, JR., San Bernardino County Department of Communications and Emergency Services

ROSS HERRICK, San Bernardino County Department of Architecture and Engineering

H. L. HOLLAND, Los Angeles Department of Water and Power

ROBERT HORRIGAN, Riverside County Emergency Operating Center

CHIEF BOB HOSKINS, Riverside City Emergency Operating Center

WALTER HOWARD, San Bernardino County Medical Center

International Conference of Building Officials

ALFRED M. KEMPER, Architect

PATRICK J. LAMA, Mason Industries, Inc.

DR. DAVID J. LEEDS, Engineering Seismologist

Liskey Architectural Mfg., Inc.

GENE MARSHALL, Photographer, Wyle Laboratories

DOM MAZZER, Riverside County Sheriff's Department

JOHN McCOY, CHIEF DEPUTY, Riverside County Sheriff's Department

JOYCE MICALLEF, San Bernardino County Department of Communications and Emergency Services

ROBERT MITTELSTADT, Architect

National Science Foundation

Oasis Water Coolers

DAVE OWEN, Wyle Laboratories

Precision Metal Products, Inc.

ROBERT REITHERMAN, Building Systems Development

Riverside General Hospital

WINTON ROSS, Jerry L. Pettis Memorial Veterans Hospital, Loma Linda

HERMAN O. RUHNAU, F.A.I.A., Ruhnau • Evans • Ruhnau • Associates

GEORGE SHIPWAY, Wyle Laboratories

ANN SKAFF, McCue-Boone-Tomsick Associates

RAY SMITH, Riverside County Department of Building and Safety

Solar Turbines International

LEON STEIN, State of California Office of the State Architect

RICHARD W. STEINMETZ, Riverside County Communications Center
Vulcan-Hart Corporation
DONALD WAYNE, San Bernardino County Data Processing Operations
LT. MARSH WHITE, Los Angeles City Police Department
PATRICK P. WONG, Los Angeles Department of Water and Power
Wyle Laboratories, Scientific Services and Systems Group

G. L. McG.

Contents

Earthquake Protection
of Essential
Building Equipment

Introduction

Earthquake resistant design concepts for the structural portions of buildings are not new to the building professions. Consideration of the equipment within a building is, however, another story. Only within the last decade have building codes addressed the survivability requirement of equipment. Building codes are just awakening to the realization that structures are being designed so as to commonly survive an earthquake with only minor damage, but that equipment generally does not fare quite so well. Recent codes are placing new demands on design professionals, manufacturers, facility owners, and governing agencies by requiring that "essential facilities" remain operational after an earthquake. The new requirements mean that more than the structural skeleton must be considered when designing for earthquake resistance.

Several studies justify these operability requirements. The California Division of Mines and Geology postulated in 1973 (C.D.M.G. Bulletin 198) that California alone could expect to lose approximately $21 billion between the years 1970 and 2000 as a direct result of earthquake shaking if current design principles are not appreciably altered. This report also postulated that with a minimum expenditure for mitigating measures, the $21 billion figure could be reduced by 50 percent. This same expenditure would also result in an approximate 90 percent reduction in loss of life. A 1977 A.I.A. Research Corporation study postulated that essential facilities such as hospitals and police stations could expect significant increases in the demand for their services immediately following a major earthquake. Unless steps are taken to prevent equipment failures wherever possible, those facilities most required for the "good of the public" after an earthquake will be unable to perform their designated tasks. Equipment failures can and do prevent the successful operation of otherwise intact facilities. Building equipment such as emergency power supplies, communications equipment, and office furniture is often unmercifully strewn about the work spaces and may be left totally inoperable after the earthquake. This scenario is not necessary. The technology exists for the seismic qualification of essential building equipment prior to the test of a major earthquake.

Building codes are generally written with minimum design requirements that are invariably applied as blanket maximums. Current seismic qualifica-

1

tion programs recommend integrated qualification procedures that apply to the individual equipment characteristics rather than blanket approaches for all equipment types.

The first two chapters of this book and Appendix 1 are intended to familiarize the reader with the nature of earthquakes, the basics of earthquake resistant design, and example building codes. This material is not intended to exhaust the respective subjects, but rather to provide information to those unfamiliar with the concepts.

Chapters 3 and 4 and Appendix 3 consume the majority of the text. It is here that the reader is introduced to a comprehensive seismic qualification plan that can be applied equally to critical equipment such as a life support system and to general support equipment such as simple shelved items. Terms and concepts such as seismic categories, earthquake testing, analysis, seismic specifications, and suggested installation details are introduced, defined, and examined. Once the reader is familiar with these concepts, a wide range of individual equipment items are discussed and suggestions are made for implementing the techniques of seismic qualification.

The principles presented herein are applicable to almost all building types. They are just as valid for office buildings, schools, refineries, and just about any place else as they are for essential facilities. The key lies in the prudent application of the concepts and diagrams of Chapters 3 and 4. It might be shown, for instance, that a computing center can justify a strict interpretation of these principles so as to reduce the potential for large economic losses due to damaged computers. These losses can generally be prevented with a few well placed bolts. A supermarket, on the other hand, may only need to employ shelf parapets to significantly reduce the potential for lost stock after every minor earthquake. It is hoped that prudent application of appropriate suggestions within this book can contribute to the reduction in the loss of life and capital throughout earthquake prone areas.

1

The Nature of
Earthquakes—Basic Principles

This discussion on earthquakes is included for those who are not familiar
with the nature of earthquakes. A basic understanding of the subject is
required to grasp the significance of this book. Those who are familiar with
the topic may wish to skip ahead to subsequent chapters.

Of the natural phenomena associated with the earth, the earthquake is one
of the most awesome and least understood. At the current time, the location,
time, and size of an earthquake cannot be accurately predicted. We can only
estimate how any particular site can be expected to shake. From the time of
the ancients right up to modern man, many explanations for the causes of
earthquakes have been given. Hopping giant toads have been blamed, as
have angry gods and collapsing gas chambers.

With modern instrumentation such as the seismograph and magnetometer,
earth scientists have been able to make intelligent guesses at earthquake
mechanisms. Through the use of the seismograph, it was determined that
the earth consists of a dense core surrounded by a mantle. The mantle is
sheathed by a thin layer called the crust. These relationships are shown in
Figure 1.1. The inner portion of the core is assumed to be solid as a result of
gravitational loading, while the outer portion is assumed to be a dense liquid.
The core is expected to be composed chiefly of iron and nickel. The mantle,
for the most part, has a plastic consistency, but it has local hot spots in its
upper region that are probably liquid. The hot spots (magma) are the derivation of the molten material of volcanoes. Surrounding the mantle is a thin
crustal layer of basaltic and granitic material. The thinnest portions (3 miles)
of the crust are in the ocean basins and are chiefly composed of basalt
(iron- and magnesium-rich silicates) that comes from the upper mantle. The
continents are the thickest portion of the crust (5–25 miles) and are composed of relatively lighter granitic (alumino silicates) rocks.

The crust is broken into what geologists call tectonic plates, which float on
the heavier mantle material like slag in a blast furnace because of their
differences in density. The plates are not in fixed positions and their motions
can be traced by geologists for approximately the last half billion years.
Current theory postulates that convection currents within the plastic mantle

3

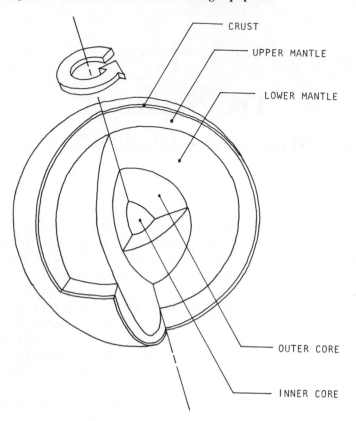

CRUST

UPPER MANTLE

LOWER MANTLE

OUTER CORE

INNER CORE

FIGURE 1.1.   The interior of the earth.

and upwellings of magmatic material at spreading centers drive the plates as if they were on a conveyor belt as shown in Figure 1.2. New crustal material is continuously being added at the spreading centers. Since the earth is not changing appreciably in size, we need to discuss the mechanism that allows new crustal material to be added at the plate boundaries and at the same time keeps the earth from growing.

Where two plates meet, an interaction occurs if they are being continuously driven. Three types of interaction are possible. First, a subduction zone can be developed when a granitic continental plate meets an oceanic basaltic plate. The lighter continental plate overrides the denser oceanic plate and the latter undergoes subduction as in Figure 1.3. The deep trench off the western coast of South America is an example of such a zone. Major earthquakes can be expected to occur in the vicinity of the subduction zone since one plate is being forced under the other.

The second possibility occurs when two continental land masses collide head on. The driving mechanism of the mantle continues to push the plates

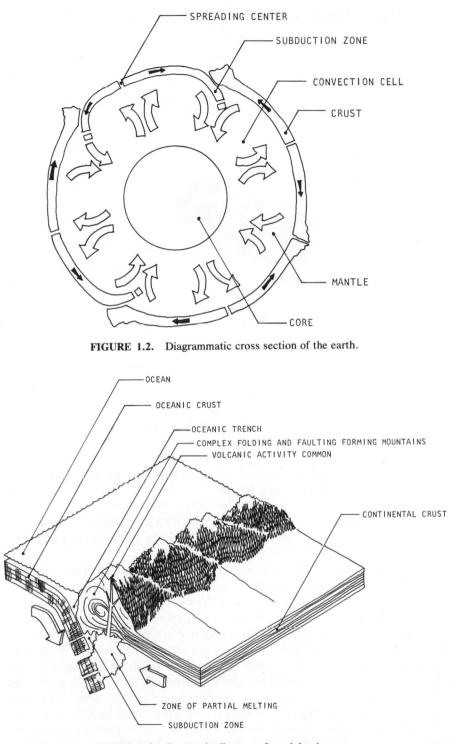

SPREADING CENTER

SUBDUCTION ZONE

CONVECTION CELL

CRUST

MANTLE

CORE

FIGURE 1.2. Diagrammatic cross section of the earth.

OCEAN

OCEANIC CRUST

OCEANIC TRENCH

COMPLEX FOLDING AND FAULTING FORMING MOUNTAINS

VOLCANIC ACTIVITY COMMON

CONTINENTAL CRUST

ZONE OF PARTIAL MELTING

SUBDUCTION ZONE

FIGURE 1.3. Isometric diagram of a subduction zone.

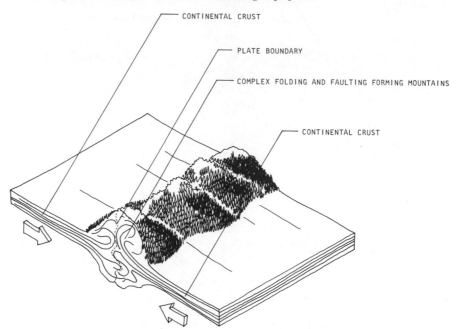

CONTINENTAL CRUST

PLATE BOUNDARY

COMPLEX FOLDING AND FAULTING FORMING MOUNTAINS

CONTINENTAL CRUST

FIGURE 1.4.   Continental plate collision.

together as shown in Figure 1.4. However, since both plates have similar densities, neither can force its way over the other and large mountains result.

The last case occurs when two plates of any density are not traveling in exactly opposite directions. Instead, they rub past one another as shown in Figure 1.5. The resulting shearing action is called strike–slip motion and is exemplified by the classic California San Andreas fault system. Everything east of the San Andreas is on the North American plate, which is moving to the west, while all that is west of the San Andreas is on the Pacific plate, which is moving to the northwest.

Observation has shown that most of the devastating earthquakes occur at plate boundaries. This is especially true at the leading edges of plates.

Figure 1.6 is an abstract projection of the earth that only intersects water. It shows where we can expect major earthquakes and volcanoes to occur. These locations principally define the major plate boundaries. Some earthquakes of various magnitudes occur, however, at places other than plate boundaries. This is the case with the western basin and range province (Sierra Nevada to Wasatch Mountains). Earthquakes occurring within the central portion of the plates are due to phenomena such as crustal thinning, magma emplacement (future batholiths), spreading center formation, or possibly old oceanic spreading centers that have been overridden by continental plates. These quakes are generally less frequent than those at plate bound-

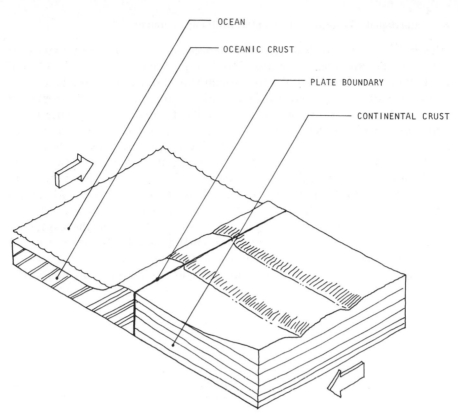

OCEAN

OCEANIC CRUST

PLATE BOUNDARY

CONTINENTAL CRUST

FIGURE 1.5. Strike–slip plate interface.

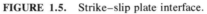
● MAJOR EARTHQUAKE ACTIVITY

FIGURE 1.6. Major earthquake occurrences mark plate boundaries.

aries but often just as destructive. Examples of interplate earthquakes are the New Madrid quakes in the Mississippi Valley, 1811 to 1812, and earthquakes associated with active volcanism in the Hawaiian Islands.

As the various plates push on each other, stresses within the rocks are developed. The rocks of the crust are similar to other building materials in that they have both yield points and ultimate strengths. If the temperature and pressure are correct, as the stress increases, the rocks behave in a plastic manner and an earthquake does not occur. Such is the case in the plastic region of the mantle. The crustal materials, however, have more of a tendency to be brittle and tend to reach their ultimate strength without any appreciable yielding. When the crust is stressed to its ultimate strength, the rocks break with a sudden stress drop. This breakage and the associated release of energy is the beginning of the earthquake. Earthquake terminology is presented in Figure 1.7. The point at which the rocks first break is termed the hypocenter or focus. The epicenter is the point at the earth's

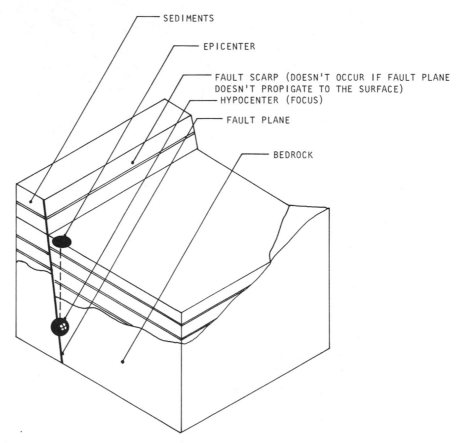

FIGURE 1.7.   Fault terminology.

surface directly above the hypocenter. The fault plane, which is the plane of weakness along which the earthquake propagates, may or may not extend to the surface. It is common, however, for the fault plane to break the surface in California. Generally speaking, the larger the plane of rupture, the greater the earthquake. For example, Dr. G. W. Housner reports in *Earthquake Engineering*, Chapter 4, that an idealized relationship between magnitude and length of fault can be drawn. This relationship seems to hold mostly for strike–slip faulting. The largest recorded earthquakes, of magnitude 8.8 or so, should have an associated rupture length of approximately 1000 miles. Smaller quakes, say magnitude 8.0, break approximately 190 miles, magnitude 6.5 quakes break 9 miles, and magnitude "0" earthquakes break approximately 100 feet.

The foregoing paragraph relies heavily on the word magnitude. The most common use of the word comes from Dr. Charles Richter, who defined the open ended logarithmic scale so that seismologists would have a method of comparing individual seismic events independently of their location (Richter, 1958).

It is a measure of the maximum single amplitude trace of the scribing instrument on a Wood–Anderson long period seismograph that is recording at a distance of 100 kilometers (62.5 miles) from the epicenter.

The seismograph is not always of this particular brand and it is obviously not always located 100 kilometers from the epicenter. Therefore, seismologists have developed mathematical relationships that allow them to measure the seismic event with their own specialized instruments. Figure 1.8 is a simplified diagram showing the principle of the seismograph. All of this is actually accomplished with sophisticated technology that can be better reviewed by the interested reader in elementary seismology texts (e.g., Richter, 1958).

Magnitude, through mathematical manipulation, can approximately describe the amount of energy released by the earthquake. Richter (1958) used the following formula (others alter it slightly, especially at its upper values):

$$\log_{10}E \text{ (ergs)} = 11.8 + 1.5M$$

The magnitude is represented by M in this formula. This results in the relationship seen in Table 1.1, which is an approximate increase of 31.5 times the energy for each full increase in magnitude. Bearing this in mind, we see that the energy released for the 6.5 magnitude 1971 San Fernando earthquake is about 1000 times less than that released in the 1964 Alaskan earthquake, which had a magnitude of approximately 8.5.

The term magnitude is actually only useful to a handful of earth scientists. A 7.0 magnitude earthquake occurring in the Marianas Trench (subduction zone) near Guam in the west Pacific Ocean at a depth of 300 miles (well within the mantle—a deep focus earthquake) is not likely to do much dam-

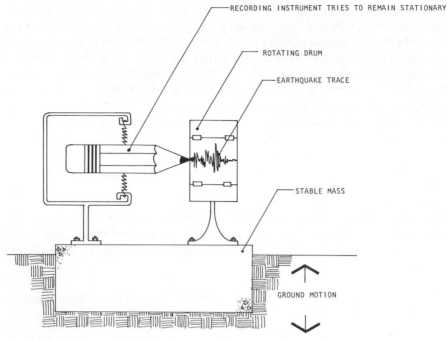

FIGURE 1.8.   Diagrammatic seismograph.

age to man-made structures. In fact, very few people would even feel it. The seismographs would record it accurately; but it would have no real significance to most people. To discuss the observed effects of the earthquake, we need yet another scale. The subjective Modified Mercalli Intensity (MMI) Scale, 1931 is reproduced in Table 1.2. This scale uses Roman numerals I to XII. It describes the effects of the earthquake at the

TABLE 1.1   Energy Values of Richter Magnitudes

Energy Released (ergs)	Richter Magnitude
2×10^{10}	-1
6.3×10^{11}	0
2×10^{13}	1
6.3×10^{14}	2
2×10^{16}	3
6.3×10^{17}	4
2×10^{19}	5
6.3×10^{20}	6
2×10^{22}	7
6.3×10^{23}	8
2×10^{25}	9

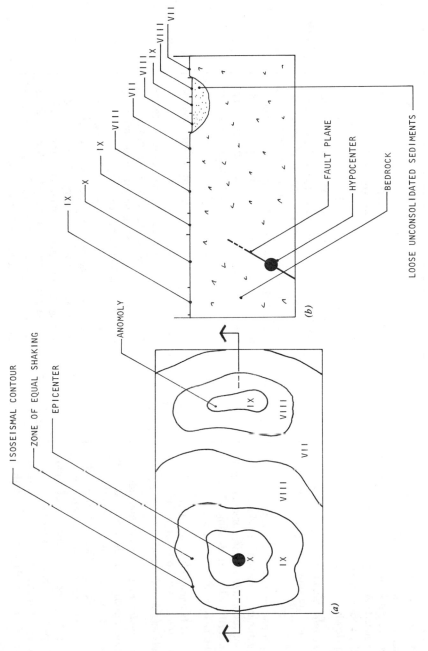

ISOSEISMAL CONTOUR

ZONE OF EQUAL SHAKING

EPICENTER

ANOMOLY

IX

X

IX

VIII

VII

VIII

VII

VIII

IX

VIII

VII

FAULT PLANE

HYPOCENTER

BEDROCK

LOOSE UNCONSOLIDATED SEDIMENTS

(b)

IX

VIII

VII

X

IX

(a)

FIGURE 1.9. Isoseismal map and cross section with anomaly.

11

TABLE 1.2 The Mercalli Intensity Scale (As Modified by Charles F. Richter in 1956 and rearranged)

If Most of These Effects Are Observed	Then the Intensity Is:
Earthquake shaking not felt. But people may observe marginal effects of large distance earthquakes without identifying these effects as earthquake-caused. Among them: trees, structures, liquids, bodies of water sway slowly, or doors swing slowly.	I
Effect on people: Shaking felt by those at rest, especially if they are indoors, and by those on upper floors.	II
Effect on people: Felt by most people indoors. Some can estimate duration of shaking. But many may not recognize shaking of building as caused by an earthquake; the shaking is like that caused by the passing of light trucks.	III
Other effects: Hanging objects swing. Structural effects: Windows or doors rattle. Wooden walls and frames creak.	IV
Effect on people: Felt by everyone indoors. Many estimate duration of shaking. But they still may not recognize it as caused by an earthquake. The shaking is like that caused by the passing of heavy trucks, though sometimes, instead, people may feel the sensation of a jolt, as if a heavy ball had struck the walls. Other effects: Hanging objects swing. Standing autos rock. Crockery clashes, dishes rattle or glasses clink. Structural effects: Doors close, open or swing. Windows rattle.	V

If Most of These Effects Are Observed	Then the Intensity Is:
Effect on people: Difficult to stand. Shaking noticed by auto drivers. Other effects: Waves on ponds; water turbid with mud. Small slides and caving in along sand or gravel banks. Large bells ring. Furniture broken. Hanging objects quiver. Structural effects: Masonry D[a] heavily damaged; Masonry C[a] damaged, partially collapses in some cases: some damage to Masonry B[a], none to Masonry A[a]. Stucco and some masonry walls fall. Chimneys, factory stacks, monuments, towers, elevated tanks twist or fall. Frame houses moved on foundations if not bolted down; loose panel walls thrown out. Decayed piling broken off.	VIII
Effect on people: General fright. People thrown to ground. Other effects: Changes in flow or temperature of springs and wells. Cracks in wet ground and on steep slopes. Steering of autos affected. Branches broken from trees. Structural effects: Masonry D[a] destroyed; Masonry C[a] heavily damaged, sometimes with complete collapse; Masonry B[a] is seriously damaged. General damage to foundations. Frame structures, if not bolted, shifted off foundations. Frames racked. Reservoirs seriously damaged. Underground pipes broken.	IX

12

Effect on people: Felt by everyone indoors and by most people outdoors. Many now estimate not only the duration of shaking but also its direction and have no doubt as to its cause. Sleepers wakened.

Other effects: Hanging objects swing. Shutters or pictures move. Pendulum clocks stop, start or change rate. Standing autos rock. Crockery clashes, dishes rattle or glasses clink. Liquids disturbed, some spilled. Small unstable objects displaced or upset.

Structural effects: Weak plaster and Masonry D[a] crack. Windows break. Doors close, open or swing.

VI

Effect on people: Felt by everyone. Many are frightened and run outdoors. People walk unsteadily.

Other effects: Small church or school bells ring. Pictures thrown off walls, knicknacks and books off shelves. Dishes or glasses broken. Furniture moved or overturned. Trees, bushes shaken visibly, or heard to rustle.

Structural effects: Masonry D[a] damaged; some cracks in Masonry C[a]. Weak chimneys break at roof line. Plaster, loose bricks, stones, tiles, cornices, unbraced parapets and architectural ornaments fall. Concrete irrigation ditches damaged.

VII

Effect on people: General panic.

Other effects: Conspicuous cracks in ground. In areas of soft ground, sand is ejected through holes and piles up into a small crater, and, in muddy areas, water fountains are formed.

Structural effects: Most masonry and frame structures destroyed along with their foundations. Some well-built wooden structures and bridges destroyed. Serious damage to dams, dikes and embankments. Railroads bent slightly.

X

Effect on people: General panic.

Other effects: Large landslides. Water thrown on banks of canals, rivers, lakes, etc. Sand and mud shifted horizontally on beaches and flat land.

Structural effects: General destruction of buildings. Underground pipelines completely out of service. Railroads bent greatly.

XI

Effect on people: General panic.

Other effects: Same as for Intensity X.

Structural effects: Damage nearly total, the ultimate catastrophe.

Other effects: Large rock masses displaced. Lines of sight and level distorted. Objects thrown into air.

XII

Source: California Division of Mines and Geology, Bull. 198, *Urban Geology Master Plan for California*, 1973.

[a] Masonry A: Good workmanship and mortar, reinforced; designed to resist lateral forces. Masonry B: good workmanship and mortar, reinforced; Masonry C: good workmanship and mortar, unreinforced; Masonry D: poor workmanship and mortar and weak materials, like adobe.

point of observation. This scale is useful in that geologists and laymen can observe shaking effects of the earthquake at their individual locations. Another use is that geologists can produce maps that are useful to the various planning professions. These so called isoseismal maps (as shown in Figure 1.9*a*) can give an indication of which areas shook the hardest during an earthquake. As the map and the cross section (Figure 1.9*b*) show, severe shaking can occur at some distance from the epicenter. Distance from the epicenter alone is not a safety factor. The pockets of higher intensity VIII and IX within the broader zone of intensity VII can represent poor soil or rock conditions.

Various professions are currently trying to adequately define a more objective intensity scale or at least to link objectivism to the MMI scale. The

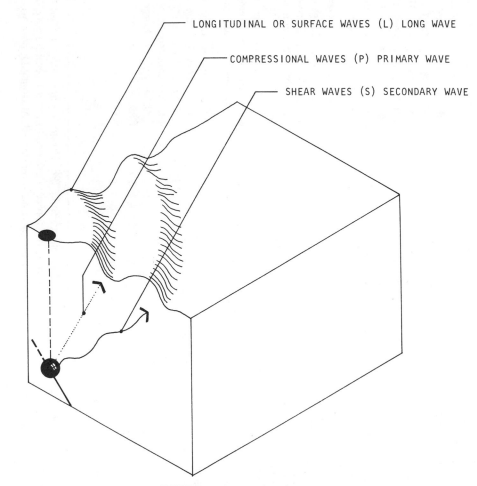

LONGITUDINAL OR SURFACE WAVES (L) LONG WAVE

COMPRESSIONAL WAVES (P) PRIMARY WAVE

SHEAR WAVES (S) SECONDARY WAVE

FIGURE 1.10.   Earthquake waves.

best results to date are noted where velocities or accelerations at the ground surface are linked to the MMI scale. More work needs to be done so that observers within buildings can also relate their experiences. There is no logical method to compare MMI to magnitude. The intensity of shaking is only indirectly related to the magnitude of the earthquake. Often, because of poor soil conditions, or distance from the epicenter or energy focusing, a smaller magnitude earthquake can produce greater shaking characteristics than a larger magnitude quake. Some individuals have produced formulas that attempt to correlate MMI and magnitude, but there are too many variables for these correlations to be of any real value.

The rocks that break during an earthquake release most of their stored energy in the form of random mechanical waves. Figure 1.10 shows the three principal types of waves that are released. The compressional waves (primary) shown in Figure 1.10 are similar to acoustic waves that travel from 15,000 to 20,000 feet/second and have ground particle displacements parallel to their direction of travel. The shear waves (secondary) shown in Figure 1.10 travel approximately 10,000 feet/second and have particle displacements normal (at right angles) to their direction of travel. The third wave is a combination of the first two. They are appropriately named the long waves. These waves basically radiate outward from the epicenter along the earth's surface similar to the concentric waves caused by a pebble thrown into a pond. The physical characteristics of these waves vary according to parameters such as rock and soil type, fault plane propagation, fault plane and lithologic geometry, topography, and soil water content. Figure 1.11 shows how the various characteristics of the subsurface geology can affect the

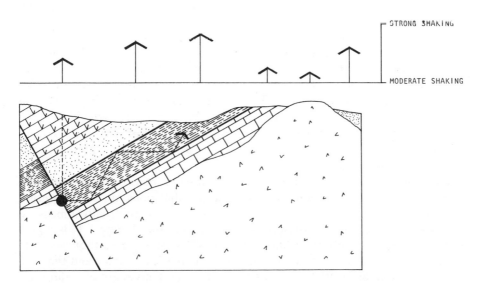

FIGURE 1.11. Subsurface effects on ground response (relative degree of shaking).

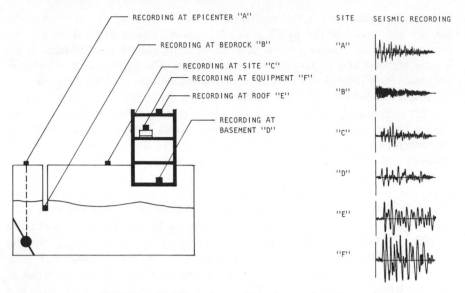

FIGURE 1.12.   Various recordings from a single event.

earthquake waves. The vertical arrows above the cross section correspond to relative intensities of shaking at various locations. The lithologic interfaces beneath the surface can develop relative shadow zones as well as focus great amounts of energy at specific sites. This latter example is the exception rather than the rule but is useful because it dramatizes the potential for complex surface effects that must be considered when assessing the earthquake problem for any given site.

Buildings and their associated components have the same potentials for complexities as the crust itself. Everything—soil, rocks, and buildings—tends to act as a filter. Some earthquake frequencies may be attenuated, while others are amplified as the waves pass from rock to soil, soil to building, and building to component. This complexity, which is diagrammed in Figure 1.12, must always be borne in mind when reviewing any earthquake resistant design. The various recording points (A through F) and their associated records show the various differences possible in wave patterns from a single earthquake event.

Seismographs and strong motion instruments record the earthquake waves. The latter instrument is a seismograph that does not record until a threshold vibration limit is reached. These recordings are usually a measure of displacement, velocity, or acceleration versus time. Figure 1.13 is a portion of a seismograph recording from the 1964 Alaskan earthquake. The seismograph continuously records displacement versus time, and the particular instrument in the figure was located at the University of California, Berkeley. Figure 1.14, on the other hand, is a strong motion recording of the

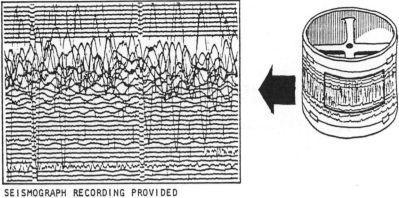

SEISMOGRAPH RECORDING PROVIDED
BY DR. HARRY COOK

FIGURE 1.13.   Portion of 1964 Alaskan earthquake record.

1940 El Centro earthquake. All three components (north—south, east—west, and vertical) of motion are shown. This particular machine recorded acceleration (% gravity) versus time. The earthquake records were terminated at 40 seconds. The strong shaking characteristics only lasted for approximately 26 seconds.

This concludes our discussion on earthquakes. How can we correlate the shaking characteristics of the earthquake to the ground, building, and finally the nonstructural components themselves? This is the topic of the next chapter, which is designed to complement this chapter by answering questions such as this one.

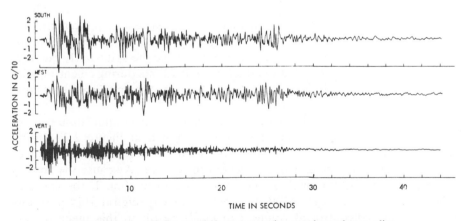

FIGURE 1.14.   El Centro 1940 strong motion accelograph recording.

2

Earthquake Resistant
Design—Basic Principles

Whether we are designing for the structure of the building or the nonstructural components located in the building, the principles of earthquake resistant design are the same. Most existing codes involved with regulating nonstructural component design require static equivalents that derive from the dynamic principles discussed below. We discuss the relevance of this practice at a later time.

This Chapter is similar to the preceding familiarization chapter on earthquakes. It is written for those who are not familiar with the dynamic principles of earthquakes, their effect on buildings, and the nonstructural components within them. The reader who does have a basic understanding of this concept may wish to skip ahead to the next chapter. Further readings may also be found in books such as Alfred M. Kemper's *Architectural Handbook*, 1979.

To illustrate the difference between what is called a static load and a dynamic load, we can draw a parallel to a carpenter driving a spike with a 2 pound sledge-hammer as shown in Figure 2.1. Just resting the hammer on the spike (the static condition) will never get it driven into the piece of wood, while the dynamic pounding of the sledge will drive the spike fairly quickly. This example can be extended to building components during an earthquake because buildings and their components behave dynamically rather than statically.

As an example, test and monitoring equipment in communication centers is commonly mounted on portable carts as shown in Figure 2.2. During normal service, the cart remains stationary on the floor through the action of the earth's gravity. The dynamic effects of the earthquake cause the floor to vibrate, which puts the building components into motion. If the cart has wheel locks and they are not set, the cart can be expected to shift positions on the floor. Uncontrolled rolling can easily result in this piece of test equipment colliding with some of the exposed relay panels, and so on. This unnecessary collision can result in extensive damage to the communication center. Setting the wheel locks, however, does not assure seismic integrity of this equipment. It only keeps the equipment from rolling in an otherwise

18

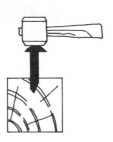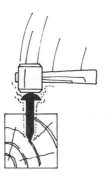

(a)   (b)

STATIC   DYNAMIC

FIGURE 2.1.   Static versus dynamic loading.

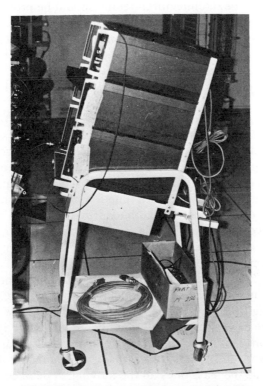

FIGURE 2.2.   Communications test equipment.

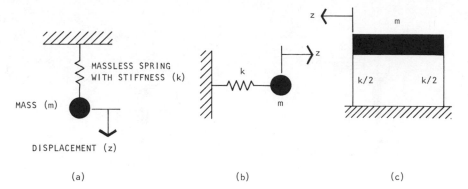

FIGURE 2.3. Typical SDOF system dynamic models without damping. (*a*) Vertical configuration. (*b*) Horizontal configuration. (*c*) With two supports.

moderate seismic environment. As the intensity of shaking increases, the portable equipment tends to wobble when the wheels are locked. If the earthquake is strong enough, the cart may skid or possibly overturn. This toppling can also damage critical equipment within the facility or the test equipment itself or may injure people in the vicinity. To adequately protect equipment of this type, architectural details should be provided for anchoring it in a fixed position when it is not in use.

The following is a discussion describing the mathematical relationships experienced by building equipment in the earthquake environment. For the sake of brevity, the discussion is limited to an undamped freely vibrating single degree of freedom system (SDOF), such as the typical dynamic model examples shown in Figure 2.3 (*a, b,* and *c*).

Damping is an inherent method of energy absorption through particle friction, wobbling connections, and so on, that would only serve to complicate our simplified model. The damping parameter must be considered in the final seismic design though, because it can have a great effect on the degree of response of an equipment item. The SDOF system is defined to have only one direction of motion (z), which is a function of time. The dynamic models use the mass (m) of the object, which means that in the real problem, where the weight is given, we must divide by the gravity constant (386 inches/second²). The mass in our example is considered to be infinitely stiff, while the column is considered to be a massless spring (k) that is a function of its length (l), its modulus of elasticity (E), and its moment of inertia (I). Figure 2.4(*a*) shows the maximum relative displacement (z) with respect to the rest position. Rocking of the system is ignored here. We know that the earthquake energy in Figure 2.4*b* arrives at the base of our idealized model from some distant seismic event. The foundation material shakes, while inertia tends to keep the mass at rest. Our simplified model shows the foundation at rest while the mass does all the moving. A detailed examination of the dynamic properties of the system would show that it does not matter if

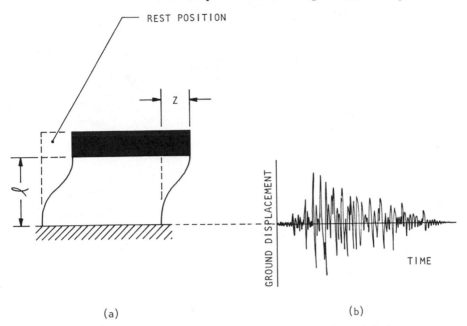

REST POSITION

z

ℓ

GROUND DISPLACEMENT

TIME

(a)

(b)

FIGURE 2.4.   Maximum displacement. (*a*) SDOF model. (*b*) Ground displacement.

we assume either the ground or the mass to vibrate, since it is the relative displacement between the mass and the ground that is significant. The assumption that the mass moves only tends to simplify the problem. Technical proofs of this assumption are available in many texts that deal with the dynamic properties of earthquakes.

We can begin to develop the required mathematical relationships by assuming that the ends of the idealized columns (which may represent a shelving unit or a communication relay panel) of Figure 2.4*a* are fixed at both top and bottom. The column shear (V) can be obtained by referring to Figure 2.5 and Equation 2.1.

$$V = \left(\frac{12EI}{l^3}\right)z \qquad (2.1)$$

The free body diagram of Figure 2.6 shows the forces applied to the mass of Figure 2.4*a*. The elastic restoring force (kz) tries to overcome the applied force [$F(t)$] and restore the mass to its original position. If we assume that a unit displacement occurs, then Equation 2.1 becomes the unitless quantity of Equation 2.2

$$V = \left(\frac{12EI}{l^3}\right)1 \equiv k \qquad (2.2)$$

which is the definition of the system stiffness (k). Formulas for arriving at k for other models can be found in elementary dynamics texts. Where more

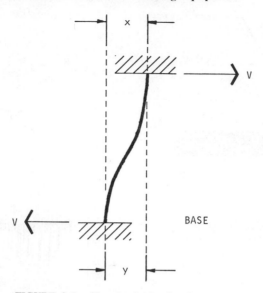

FIGURE 2.5.   Shearing in fixed end column.

than one column is considered as in Figure 2.3c, the column stiffnesses are summed.

Newton's law states that the sum of the applied force in Figure 2.6 is equal to the mass times the acceleration, or

$$\Sigma F = ma \tag{2.3}$$

where the acceleration (a) is in units of distance per second per second. The dot ($\cdot$) format is used to represent all time derivatives. With it, we can define the following relationships:

$x(t) \equiv$ the absolute displacement of the mass (distance)

$\dot{x}(t) \equiv \dfrac{dx(t)}{dt} =$ the absolute velocity of the mass (distance/second)

$\ddot{x}(t) \equiv \dfrac{d^2x(t)}{dt^2} =$ the absolute acceleration of the mass (distance/second^2)

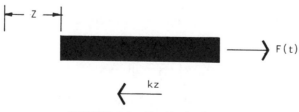

FIGURE 2.6.   Free body diagram.

Figure 2.5 shows the variables x to be the absolute displacement of the mass and y to be the absolute displacement of the ground. Figure 2.4a shows the variable z to be the maximum relative displacement or

$$z = x - y \qquad (2.4)$$

Newton's law can be applied to the mass in Figure 2.4a. The dynamic equation of motion for an undamped SDOF system is expressed in Equation 2.5.

$$F(t) - k(x - y) = m\ddot{x} \qquad (2.5)$$

By subtracting $m\ddot{y}$ from both sides of Equation 2.5, it becomes $F(t) - m\ddot{y} = m\ddot{z} = kz$. For free vibration, $F(t) =$ equals 0, and y equals 0. If we next divide both sides of Equation 2.5 by the mass m we have:

$$z(t) = \frac{k}{m}z(t) = 0 \qquad (2.6)$$

We can now define the undamped natural frequency (ω_n) of our SDOF system in radians/second as represented in Equation 2.7.

$$\omega_n = \sqrt{\frac{k}{m}} \text{ radians/second}, \qquad (2.7)$$

Next we define the undamped natural frequency (f_n) in cycles/second, or more conventionally, hertz, as shown in Equation 2.8:

$$f_n = \frac{\omega_n}{2\pi}(\text{cycles/second} = \text{hertz}) \qquad (2.8)$$

Finally, we can define the undamped natural period (T_n) in seconds as shown in Equation 2.9:

$$T_n = \frac{2\pi}{\omega_n} - \frac{1}{f_n} \text{ (seconds)} \qquad (2.9)$$

The undamped natural frequency can be arrived at directly from the weight (W) of the system as shown in Equation 2.10:

$$f_n = \sqrt{\frac{9.8k}{W}} \qquad (2.10)$$

This saves in the conversion from weight to mass.

   These relationships are important to the seismic qualification of the building equipment. We have learned that the initial earthquake generates random wave patterns. The random vibrations contain potentially destructive frequencies from about 0.5 to possibly 40 hertz as shown in Figure 2.7. This mechanical spectrum is not to be confused with the electromagnetic spectrum (light, radio waves, etc.), which is also measured in hertz. As a general rule, the lower frequencies in the range of 0.5 to 10 hertz (with higher

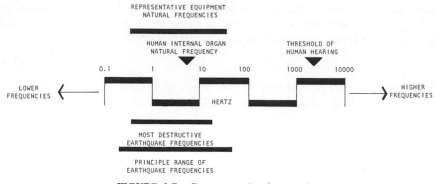

FIGURE 2.7.   Representative frequencies.

amplitudes associated) cause more problems to building equipment. Every material through which the earthquake wave travels affects the wave. This is true of the various geologic frameworks, as well as the building structure and even the equipment. The effects tend to attenuate some of the system response to the frequency content and to amplify others. By use of terms that are discussed shortly, such as ground response spectra and floor response spectra, it is possible to predict with some degree of confidence what the seismic frequency distribution at each equipment location within the structure will be. A knowledge of the expected seismic frequency distribution will tell us what associated equipment natural periods are critical. When seismic frequency distribution overlays equipment natural period, extreme equipment oscillations usually occur. This is similar to the slight push required at the correct timing intervals for a child on a swing to attain the highest arc. The more motion the equipment experiences, the more likely that the system will have a failure. The failure can result from internal component malfunctions, from overturning of the equipment, from hammering against a wall or from other pieces of equipment colliding with it.

The concept of the response spectrum has been extremely valuable to designers of earthquake resistant materials. It is developed as in Figure 2.8 from a plot of the maximum response to a SDOF system for a particular earthquake with variable frequency and variable damping.

The response spectrum is applicable to the free field environment (ground response spectrum), the building (structure response spectrum), and the equipment within the building (equipment response spectrum). The particular format most often chosen to represent the response spectrum is a four coordinate logarithmic plot. By referring to Figure 2.9, we can see these coordinates. They are the frequency (the multiplicative inverse of the period), the spectral displacement (S_d), the spectral velocity (S_v), and the spectral acceleration (S_a). The smoothed curves in Figure 2.9 are for $\frac{1}{2}$ and 5 percent critical damping. One hundred percent critical damping is the

m = Mass

δ = Damping

k = Stiffness

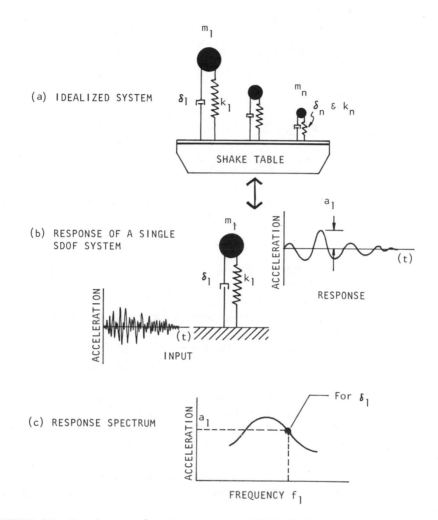

(a) IDEALIZED SYSTEM

SHAKE TABLE

(b) RESPONSE OF A SINGLE
    SDOF SYSTEM

ACCELERATION

ACCELERATION

(t)

RESPONSE

(t)

INPUT

(c) RESPONSE SPECTRUM

ACCELERATION

For δ_1

FREQUENCY f_1

FIGURE 2.8. Development of response spectrum. (*a*) Idealized system. (*b*) Response of a single SDOF system. (*c*) Response spectrum.

25

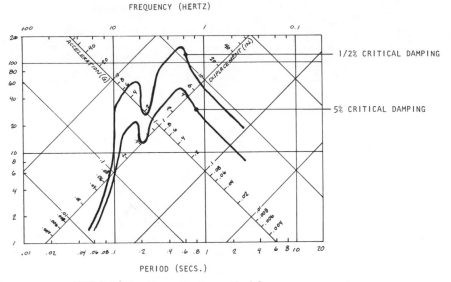

FREQUENCY (HERTZ)

1/2% CRITICAL DAMPING

5% CRITICAL DAMPING

PERIOD (SECS.)

FIGURE 2.9. Example of smoothed floor response spectrum.

amount of energy dissipation required to bring a freely vibrating system to a
halt after one full oscillation.

The response spectrum has definite limitations. It can tell us the maximum
response to a SDOF array, the peak amplitude of input, and the bandwidth
of the input. It cannot tell us the duration of the earthquake (that must come
from the strong motion recording), the type of wave of the input (in natural
earthquakes, we can assume random), how a particular piece of equipment
will respond or if a particular piece of equipment will continue to operate
during and after the earthquake. The use of the response spectrum is only as
good as the assumptions made in the initial math model.

The spectral velocity (S_v) depends on the particular characteristics of the
earthquake under consideration, the natural frequency of the system, and
the damping characteristics. The spectral displacement (S_d) and the spectral
acceleration (S_a) are associated with the spectral velocity as shown in
Equations 2.11 and 2.12. Damping has again been ignored for the sake of
simplicity.

$$S_d = \left(\frac{T_\eta}{2\pi}\right)S_v \tag{2.11}$$

$$S_a = \left(\frac{2\pi}{T_\eta}\right)S_v \tag{2.12}$$

The maximum base shear (V_b) of a piece of equipment can be obtained
from the relationship shown in Equation 2.13. The spectral velocity must be

FIGURE 2.10.   Communications rack.

derived from a specific design earthquake response spectrum such as Figure 2.9. The maximum base

$$V_b = kS_d = k\left(\frac{T_\eta}{2\pi}\right)S_v \qquad (2.13)$$

shears can also be obtained from the 1979 *Uniform Building Code* relationship shown in Equation 2.14:

$$V_b = ZIC_pW_p \qquad (2.14)$$

where Z = seismic zone coefficient
$\quad\quad I$ = building importance factor
$\quad\quad C_p$ = equipment horizontal force coefficient
$\quad\quad W_p$ = weight of the equipment

To compare these two methods, we take as a simple example a communication relay panel located in an emergency operation center. The piece of equipment shown in Figure 2.10 is similar to that which we are considering.

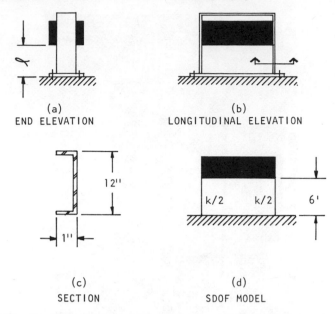

FIGURE 2.11.   Example equipment. (*a*) End elevation. (*b*) Longitudinal elevation. (*c*) Section. (*d*) SDOF model.

The actual configuration of the equipment in our example is shown in Figure 2.11 (*a–d*).

The following information is given with respect to Figure 2.11:

Z = 1.0 (seismic zone 4)
I = 1.5 (essential facility: emergency operating center)
C_p = 0.3 (equipment required for essential facility operation)
W_p = 1200 pounds

Calculate for both 0 and 5 percent critical damping in the longitudinal direction only. Calculating the *Uniform Building Code* value first, we have

$$V_b = ZIC_pW_p = (1) \times (1.5) \times (0.3) \times (1200)$$

$$V_b = 540 \text{ pounds}$$

The *Uniform Building Code* method does not take into account the dynamic properties of the system. To do this, we use the response spectrum approach and make a simple dynamic analysis for our idealized SDOF equipment system.

Determining the stiffness for the columns, the natural frequency, and natural period of the system:

$$k = \frac{12EI^*}{l^3} = \frac{(12) (29 \times 10^6) (0.015)}{(6 \times 12)^3} = \frac{5.2 \times 10^6}{72^3} = 139$$

*Not *UBC I*.

$$\omega_\eta = \sqrt{\frac{k_1 + k_2}{m}} = \sqrt{\frac{(139 + 139) \, 386}{1200}} = 9.48 \text{ radians/second}$$

$$f_\eta = \frac{\omega_\eta}{2\pi} = 1.5 \text{ hertz}$$

$$T_\eta = \frac{2\pi}{\omega_\eta} = 0.66 \text{ seconds}$$

Referring to Figure 2.9, we find that the spectral velocity is approximately 90 inches/second for $\frac{1}{2}$ percent critical damping and 38 inches/second for 5 percent critical damping. The base shear Equation 2.13 is then:

$$V_b = k \left(\frac{T_\eta}{2\pi} S_v\right)_{1/2\%} = 278 \left(\frac{0.66}{2\pi} \times 90\right) = 2630 \text{ pounds}$$

$V_b = 2630$ pounds for $\frac{1}{2}$ percent critical damping

$V_b = 1110$ pounds for 5 percent critical damping (calculations omitted)

Thus we can see that the *Uniform Building Code* base shear value is significantly less than that of the response spectrum solution. Neither of these methods, however, can confirm that equipment will remain operational during and after an earthquake. To assure equipment operability for most items during and after an earthquake, it must be subjected to a dynamic testing procedure (earthquake simulations).

Above we discuss only single degree of freedom systems. Many of the building components encountered by the design professional are more complex in their dynamic response to an earthquake. These complex systems are commonly termed multiple degree of freedom (MDOF) systems. Figure 2.12 shows an example of a typical MDOF system that needs to be given seismic consideration. This type of equipment item is commonly required for the effective operation of facilities after an earthquake.

There is more than one method to determine the associated forces and response characteristics of the MDOF system. We discussed the response

FIGURE 2.12. Multiple degree of freedom example.

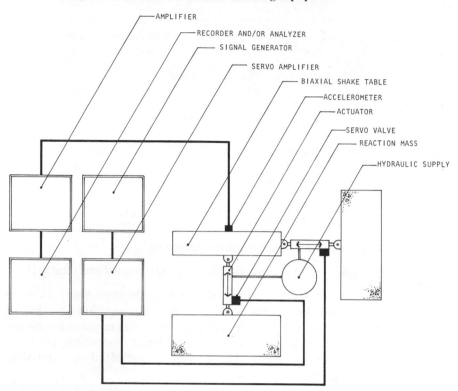

FIGURE 2.13.   Diagrammatic biaxial seismic simulator.

spectrum method for the SDOF system earlier. The principal difference in calculating for the MDOF as opposed to the SDOF is the number of natural frequencies (a dynamic system having the same number of natural frequencies and associated modes of vibration as there are degrees of freedom) and the physical method of analyzing the math model. The response spectrum method of analysis usually requires the least expenditure of engineering and computer time.

The other principal method of analysis is called the time-history method. This method is complicated, requires extensive computer application, and may be beyond the scope of the architectural profession. This, however, does not preclude the architect from being aware that this method exists. The time-history method yields a more exact approximation of the expected behavior of the MDOF system. Where the response spectrum method yields results that are usually conservative, the time-history method of analysis yields results that take into account each increment of time for the seismic event, thus providing results that are generally more indicative of the true system response.

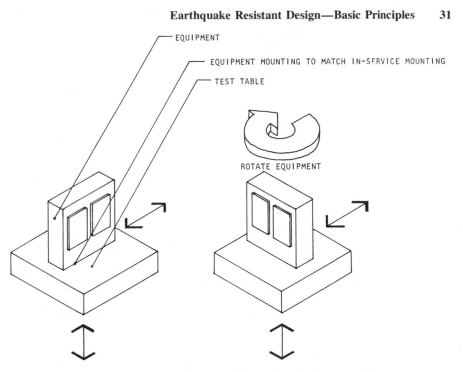

EQUIPMENT

EQUIPMENT MOUNTING TO MATCH IN-SERVICE MOUNTING

TEST TABLE

ROTATE EQUIPMENT

FIGURE 2.14. Equipment rotation on the seismic test table.

There are specific pieces of equipment that must remain functional after an earthquake for the safe and efficient operation of an essential facility. A seismic analysis cannot always guarantee that this equipment will operate. When this is the case, an earthquake test on a seismic simulator is required. Because of equipment limitations, it is sometimes necessary to combine the test procedure with an analysis. Figure 2.13 is a schematic drawing of a seismic test machine.

The nonstructural component under consideration is mounted on the test table in a fashion similar to that of the actual mounting in the essential facility. During and after the earthquake test, the equipment is subjected to functional tests. This is principally the only method of assuring that the equipment will be able to function after an earthquake.

A number of options are available with the earthquake test. First is the choice of the number of axes to be tested. Current technology limits seismic testing to only one or two axes of excitation. Biaxial simulators usually have one horizontal component and the vertical component operating simultaneously. To test the two equipment plan axes (transverse and longitudinal), the test specimen is rotated between tests as in Figure 2.14. The next major decision with respect to seismic testing is the type of waveform to be used.

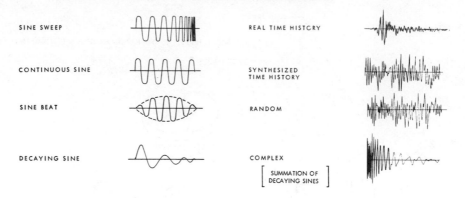

FIGURE 2.15.   Waveforms available for testing.

Figure 2.15 shows the eight basic waveforms that are available for testing purposes. Each of these waveforms has its own particular use when applied to the test specimen. For instance, the sine sweep is used to determine the specimen's natural frequency. The sine dwell test can be used to determine failure characteristics of the specimen. The time histories and random waveforms are used most often for analysis verification and proof testing. Proof testing is the most valuable asset of the seismic test philosophy. This is the method by which a particular piece of critical equipment can be tested to relatively guarantee that it will be able to perform its function after an earthquake. The equipment is subjected to its required operational functions during and/or after the earthquake test to be certain that it will perform adequately.

This concludes our discussion on the basic principles associated with earthquake resistant design. The next two chapters present and discuss fairly specific details on the qualification approaches and procedures for many types of equipment found in modern buildings.

3
Essential Facilities, Qualification Programs, Systems, and Equipment

Essential facilities are the backbone of emergency efforts immediately after an earthquake. Their continued and successful operation is of the utmost importance for the community's well-being.

Early use of the essential facility concept came with the Civil-Defense Emergency Operating Centers. Then in 1975, the Structural Engineers Association of California (SEAOC) published recommendations for lateral force design and defined "essential facilities." The 1976 *Uniform Building Code* adopted SEAOC's definition in Paragraph 2312, Section (k), which reads:

> **2312(k) ESSENTIAL FACILITIES.** Essential facilities are those structures or buildings which must be safe or usable for emergency purposes after an earthquake in order to preserve the health and safety of the general public. Such facilities shall include, but not be limited to:
> 1. Hospitals and other medical facilities having surgery or medical treatment areas.
> 2. Fire and Police Stations.
> 3. Municipal Government Disaster Operation and Communication Centers deemed to be vital in emergencies.

Some governing bodies have decided that facilities such as computing centers and general services buildings also fall into the essential facility category even though they are not specified in the building codes. In the case of computing centers, substantial savings may be realized by providing seismic protection programs for the survivability of the equipment. Expensive and sensitive computer equipment if correctly protected will not be as likely to be damaged. Equipment that is damaged must be repaired or replaced, often at very high cost. Relatively inexpensive protection measures taken in the design phase may reap substantial savings when a damaging earthquake strikes. Prior determination of potential damage to equipment

FIGURE 3.1. Wire retainer allows for easy removal of shelved items. Photograph courtesy of Robert Reitherman.

and seismic risk dictates to the design team just when and at what levels seismic protection is needed.

Equipment seismic qualification is not limited to essential facilities by this book. Retailers, for instance, have used and can continue to use recommendations for shelved items. A simple shelf addition such as a parapet or wire retainer as shown in Figure 3.1 allows the easy removal of shelf contents and yet restrains the items during an earthquake.

DESIGN EARTHQUAKES

Chapter 2 is titled "Earthquake Resistant Design," not "Earthquake Proof Design," for very specific reasons. It is beyond current comprehension and economic feasibility to earthquake proof much of anything. Responsible designers have long realized this and have approached the problem accordingly.

Prior to addressing the seismic qualification program for the facility in question, the design team needs to assess the regional seismicity. Figure 3.2a is a seismic map showing expected qualitative seismic risk values for the United States, Canada, and Mexico. This map has been adapted and compiled from the sources indicated in the figure. Figure 3.2b is a seismic risk map showing expected bedrock-acceleration values for the contiguous United States. This map was developed by Algermissen and Perkins in 1976.

Reference to these maps during design projects provides designers with a general base of information sufficient to begin assessing the seismic risk problem. As extreme examples, designers in Milam County, Texas (the central part of the state), would not need to overly concern themselves with earthquake design for building equipment. Designers of essential facilities in San Benito, California (south of the Bay Area), on the other hand, can expect maximum potential damage and therefore earthquake resistance of building equipment should be a major concern.

For essential facility projects located in California, the designers can begin by referring to the local city or county Seismic Safety and Public Safety

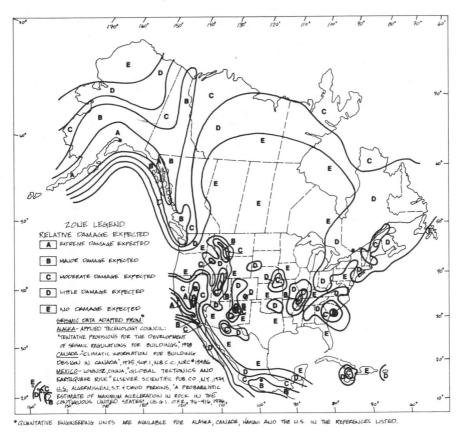

FIGURE 3.2a.   Relative seismic risk zone map—expected damage.

Elements. These documents are available through the local planning departments and describe in general terms the maximum credible and maximum expected earthquakes for the region. They are the rough beginnings of what can be termed the design earthquake. This information is generally more detailed than that found in Figure 3.2 and provides the grounds for preliminary conceptual designs prior to the actual determination of the design earthquake for the specific site.

It should be noted that Figure 3.2b designates the expected bedrock accelerations. Building and equipment response probably will be significantly different from the expected bedrock responses shown. The variables leading to these different responses include:

- Depth of soil.
- Depth and condition of sediments (consolidated versus unconsolidated).
- Groundwater content.

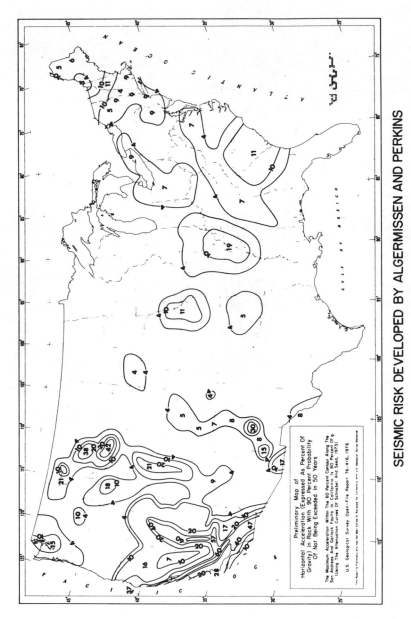

SEISMIC RISK DEVELOPED BY ALGERMISSEN AND PERKINS

FIGURE 3.2b. Seismic risk developed by Algermissen and Perkins, 1976. U.S.G.S. Open File Report 76-416, 1976. Reproduced from NBS Special Publication 510, ATC Publication ATC 3-06, NSF Publication 78-8.

- Building shape.
- Building structural type.
- Shear building.
- Moment resisting.
- Space frame.
- Number of floors.
- Equipment rigidity.
- Equipment location.
- Equipment connections.

To adequately define how a building, and subsequently the equipment, will respond, these variables must be considered when selecting or deriving the design earthquake.

To accurately predict equipment response, with any degree of certainty, the free-field design earthquake should be applied to the building so that floor responses at the equipment locations can be determined. Derivation of the floor response can be arrived at by using a step-by-step time-history analysis of the building structure. This is a complex solution and referral to structural engineers versed in its usage is suggested.

Design earthquakes are generally presented in the response spectrum or time-history format. Figures 3.3.*a* and 3.3*b* are examples of such design earthquakes in the response spectrum format. Figure 3.3*a* is the "classical" 1940 El Centro earthquake. The other is a smoothed synthesized design earthquake. The reader will note that the El Centro plot contains five curves. These various curves represent selected percentage levels of critical damping. The synthesized design earthquake contains two smoothed curves. The term "maximum credible earthquake" is the largest earthquake expected at the site. The lower curve is the "maximum probable earthquake" and represents the earthquake that is likely to occur during the useful lifetime of the facility. The foregoing terminology generally refers to the free-field earthquakes.

Various industry standards have defined their own design earthquakes for equipment:

- Safe Shut-Down Earthquake (SSE).
- Design Basis Earthquake (DBE).
- Operating Basis Earthquake (OBE).
- Design Contingency Earthquake (DCE).
- Design Operating Earthquake (DOE).

The different titles are somewhat indicative of the industry from which they arise. The first three have their genesis in the nuclear power industry, while

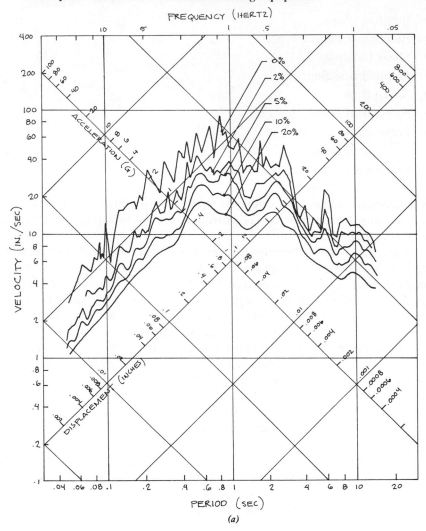

FIGURE 3.3a. Hypothetical response spectrum.

the last two were developed for the Trans-Alaska pipeline (Anderson and Nyman, 1977). Their philosophy, however, remains the same. These design earthquakes all allow the design team to compare the structure or component with that of an earthquake that is likely to occur for the particular site.

In the case of essential facilities, some equipment is more critical to the functional aspects of the facility than others. Thus it is logical to define two design levels: one aimed at critical equipment and the other aimed at support equipment. This book utilizes two such design earthquakes:

● Critical Equipment Earthquake (CEE).
● Support Equipment Earthquake (SEE).

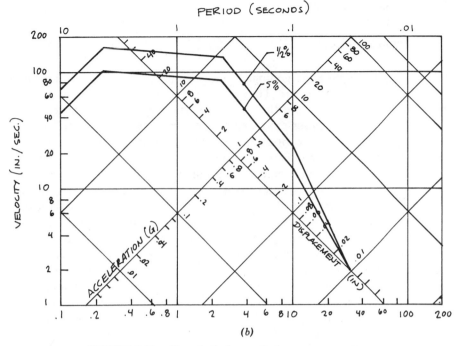

FIGURE 3.3b.   Hypothetical smoothed response spectrum.

Both of these design earthquakes are expressed originally as functions of ground motion at the particular site. The characteristics of this dual-level approach are presented below:

- The Critical Equipment Earthquake is the maximum earthquake that can reasonably be expected to occur based on a consideration of regional tectonics within the known existing geologic framework. This earthquake has a very low probability of being exceeded during the life of the essential facility.
- The Support Equipment Earthquake is the largest earthquake that the facility is likely to experience while it is in service. This earthquake has a recurrence interval from 25 to 75 years. It is possible that the facility may experience more than one SEE during its lifetime.

Formulation of the design earthquake requires interaction of several professions such as:

- Architects.
- Civil engineers.
- Engineering geologists.
- Geologists.

- Mechanical engineers.
- Seismologists.
- Structural engineers.

The geologist, engineering geologist, and seismologist interact and prepare the initial geotechnical report. They examine the regional and local geology, the tectonic framework, and the local subsurface geology. From their field observations and computer modeling they estimate the recurrence intervals and magnitudes for the various faults that could affect the facility to be designed. This information is used to postulate the expected ground motion at the site through computer models of ground interactions as the earthquake waves travel from the hypocenter to the site. The geotechnical report usually presents this information in the response spectrum or time history format.

The architect, civil engineer, structural engineer, and mechanical engineer have the task of transforming the ground response motion into the floor response motion. The earthquake waves undergo transform functions as they pass from soil to structure and on through the structure from level to level. These transform functions are based on the foundation characteristics, building structure characteristics, building geometry, damping, and so on. The derived floor response or equipment location response is also most often presented in the response spectrum format. The anticipated floor response for both the CEE and SEE is the environment for which the facility equipment must be designed.

Whether the design team applies the CEE or SEE depends on whether the equipment involved is considered to be critical or not. The CEE (higher level) is intended for use on equipment that must operate during and after an earthquake or where it is required for life support. The lower level SEE is intended for use on ancillary support equipment that is required only for day-to-day operation of the facility.

SEISMIC QUALIFICATION METHODS

The five basic methods by which a piece of equipment may be seismically qualified are

- Seismic test.
- Mathematical analysis.
- Past experience.
- Design team judgment.
- Combined qualification approaches.

The following discussion describes in detail each of these methods.

Seismic Testing

Dynamic seismic testing should be considered as the primary avenue to qualification when the equipment will be required to remain operational before, during, and after the earthquake (seismic category "A" as discussed in the next section) or if it is too complicated for an adequate mathematical analysis. Figure 3.4 is a photograph of a value subjected to a biaxial earthquake test on one of Wyle Laboratories' seismic test machines at its Norco, California facility.

The objective of seismic testing is to provide the dynamic earthquake motions in the laboratory situation that simulate the predicted building floor motion as defined by the response spectrum supplied in the design requirements. This spectrum is called the required response spectrum (RRS). The seismic test table response spectrum is termed the test response spectrum (TRS) and must envelop the RRS to be considered a valid test. The two principal types of seismic testing are proof testing and fragility testing. Proof testing is designed for equipment that meets one of the following conditions:

● The equipment is to be tested to a specific requirement for a specific application.
● The equipment is likely to require retesting for each application.

Fragility testing determines the maximum capability of the equipment for both single and multiple frequency waveforms in a manner that can be used to show compliance with future requirements. Fragility tests are conducted until the equipment fails through one mechanism or another. Malfunction (operational failure) criteria are easy to confirm by fragility test methods. Structural failures may be slightly more difficult to monitor and require a test procedure specifically designed for each individual piece of equipment. Fragility testing is an approach that equipment manufacturers may wish to

FIGURE 3.4. Staged time lapse of a valve undergoing a biaxial seismic test. Photo courtesy of Wyle Laboratories Scientific Services and Systems Group, Robert "Gene" Marshall, photographer.

take in order to supply equipment to more than one facility in an earthquake prone area.

When the design team specifies dynamic testing, they are required to provide the testing laboratory with a test specification, which should include the following information and requirements:*

● Description of in-service mounting conditions. The seismic test shall simulate these conditions.
● Description of equipment orientation.
● Axial test requirements (single axis or biaxial). Incoherent phasing between axes shall be ensured by employing separate driving signals for each axis.
● Required response spectrum (RRS).
● The test response spectrum (TRS) will envelop the RRS.
● The test laboratory shall be consulted for applicable waveforms.
● Data obtained from all the specimen response accelerometers as well as from the table response accelerometers shall first be recorded on analog tape, thus allowing the subsequent analysis and plotting of any required accelerometer record. In this case, all the table response data shall be analyzed and plotted.
● Required duration of strong motion for the earthquake simulation.
● A resonance search shall be conducted by a low level uniaxial sine sweep frequency test in each major equipment axis from 0.5 to 30 hertz. A sweep rate of 1 octave/minute shall be employed. All data shall be recorded on an oscillograph recorder.
● Accelerometers shall be mounted on both the test table and the test specimen.
● Description of the parameters that are to be examined (i.e., failure criteria, etc.).
● Specimen operational characteristics shall be monitored and recorded before, during, and after the dynamic testing program.
● Results of the dynamic testing program shall be prepared in a written report format. The report shall be certified by a registered professional engineer and include, but not be limited to, the following information:
  • Test levels.
  • Details of deviations.
  • Anomalies.
  • Repairs.

*The author wishes to express his gratitude for the assistance provided by Wyle Laboratories in the development of the following.

- Photographs of test setups and failures.
- Test log.
- Equipment listing.

Many problems with equipment have been encountered in past earthquakes. The last section of this chapter, beginning with Figure 3.146, contains photographs of examples of past failures. Dynamic seismic testing can reduce the failure potential by alerting the product designers of equipment weaknesses prior to installation. Several points have been excerpted below from an unpublished work by C. W. Roberts of Wyle Laboratories entitled *"Environmental Simulation—A Powerful Tool for the Product Designer."* The paper further emphasizes the case for dynamic testing to demonstrate operability of equipment and its value as a viable method for product design.

- Designers often soft mount equipment on vibration isolators to protect it against plant induced vibrations (or vice versa) without consideration for (motion restraint). Unfortunately, those isolators capable of performing the best job against the relative high frequency plant induced vibration will often amplify the effect of low frequency earthquakes. Dynamic seismic testing can check these mounting systems as well as the equipment response to assess survivability.
- Plug-in devices (such as connectors and printed circuit boards) vibrate loose. These items should be designed with some type of positive latch. Dynamic seismic testing can alert the product designer to potential problems before they occur.
- Easy-access type design features, such as roll-out racks, magnetic door latches, etc., will open or unlatch during an earthquake and should be replaced with positive latching features.
- Devices which are obviously sensitive to the dynamic environment such as mercury wetted switches, lightly sprung relays, etc., should not be used in earthquake applications. If they are used, they should be mounted in a manner so as to minimize the amplification of earthquake effects (i.e., at the bottom of the cabinet). Dynamic seismic testing can alert the product designer to potential problems.
- Many users will have seismic design specifications that require cabinets to be designed with no resonant frequencies below 33 hertz. Determining resonant frequencies analytically, even for relatively simple cabinets is a costly task. It is costly both economically and in time and effort expended. In the laboratory, a prototype cabinet with masses installed to simulate equipment, can be instrumented and subjected to a low-amplitude sine-sweep excitation. Resonant frequencies are very easily and accurately detected. Design changes to eliminate undesirable resonances or damping characteristics such as stiffeners, additions of foam rubber strips to door

frames, etc., can be more readily conceived in the "three dimensional" environment of prototype hardware and the "real world" atmosphere of simulated dynamic conditions.

- The typical cabinet structure is not designed with earthquake loads in mind. Panel faces are commonly attached with either self-tapping screws (which back out) or tack welds (which break). Cabinet anchoring provisions are commonly inadequate. Cabinets should be constructed using through bolts and locknuts or at least 25 percent welds.

- Many equipment cabinet designers will locate heavy equipment at the top of the cabinet for functional purposes. An example is shown in Figure 2.10. This "top heavy" arrangement has been observed to produce large dynamic moments, excessive deflections and high stresses during earthquake simulations.

As useful as seismic testing is, it should not be employed for all qualification projects. Seismic testing will prove to be a valuable research tool for examinations of equipment. In practice, however, it should be used where the results of the test provide the best information for the money expended on qualification.

Recent draft legislation in a state with a recognized seismic probability would have had manufacturers testing virtually all the equipment that they would be selling to certain essential facilities for base anchorage. This is an inappropriate use of testing and would have been analogous to using a bulldozer for planting a rose bush. Where base anchorage is the only concern, the qualification program can be best completed by mathematical analysis, which is discussed below.

Mathematical Analysis

A mathematical analysis may be applicable to all the seismic categories from "A" to "E," which are discussed in the next section, depending on the nature of the equipment and its intended use. Analysis is the preferred method of seismic qualification when the following criteria can be met:

- The equipment is required only to operate after an earthquake (i.e., the only concern during the earthquake is its structural integrity).
- The equipment can be mathematically modeled.

Examples of required mathematical analyses are shown in Table 3.1 for representative equipment conditions. Mathematical analyses are often not recommended for complex equipment. Several decisions must be made when equipment is considered for analysis. Is a simple static analysis to be used or a dynamic analysis? The simple static analysis is appropriate where equipment only need be adequately anchored. More complex installations of equipment require the construction of a dynamic math model for further

TABLE 3.1.   Mathematical Analysis Required

If the Equipment is:	Analyze for:	By Using:
Rigid with high center of gravity	Overturning	Simple overturning and pull-out conditions of anchorage system
Rigid with low center of gravity	Shear failure of anchors	Static coefficient method for shear of anchorage system
Flexible equipment (simple)	Natural frequency, velocity, displacement and acceleration, and base anchorage	Simple dynamic math model response spectrum and base shear method or static coefficient method
Flexible equipment (complex)	Natural frequency, velocity, acceleration, and base shear	Complex dynamic math model, response spectrum, computer solution, base shear

consideration. Using the dynamic model, one can calculate the resonant frequencies. If the equipment in question is rigid, the response equals the zero period acceleration of the response spectrum. If the equipment is flexible, it is necessary to compute the dynamic response of the equipment. Rigid equipment can generally be defined as equipment with a natural frequency in the high frequency range (say more than 20 hertz). This is termed the "zero period acceleration" and is the maximum acceleration contained in a response spectrum. Flexible equipment is equipment with a fundamental frequency in the low frequency range (say less than 20 hertz). Computing the dynamic response for both rigid and flexible equipment can be conducted by means of a modal analysis. An example of such an analysis was made in conjunction with Figure 2.11.

An equipment stress analysis is the final step required for the mathematical analysis track. At this point, the designer is in the position of declaring the equipment to be qualified or not qualified. If the equipment meets all the design requirements, the designer should provide adequate documentation. If the equipment does not meet the design requirements, it should be reconsidered.

Past Experience

In many cases equipment can be qualified by past experience. The value of this qualification approach is realized as an economic saving to the facility owner and a time saving to the design team. To be valid, the past experience qualification procedure must be adequately documented and fully approved by the governing agency, the design team, and the facility owner.

Dispersal of the seismic qualification information for future projects could be best performed by a computer network. Computer companies currently have existing systems on line that could be adapted to meet agency requirements for documentation. The computer companies could sell time-sharing to the design teams, who would then be able to assess which equipment had been previously qualified for particular environments. Equipment manufacturers would be prompted to provide the computer data base with their most up-to-date qualified equipment in order to remain competitive. The computer system would be capable of providing the design team with fast, inexpensive information.

This approach is not without precedent. A similar auditing system provided the seismic qualification records for the Alaska Pipeline (Anderson and Nyman, 1977). Although these records were not intended for public dispersal, the computer programs easily could have been adapted for this function. Similar proposals have been suggested for the nuclear power industry.

In setting up such a network, the computer company would work directly with the governing agencies, selected architects and engineers, and representative equipment manufacturers. They would all need to collaborate to establish a usable data base. Some of the information that may be useful to such a program is:

- Manufacturer name.
- Equipment name, model number, and so forth.
- System of which the equipment is a part.
- Seismic qualification status.
  - Methods of qualification.
  - Qualified by whom.
  - Design earthquake utilized, and so forth.

Further examination of this proposal will indicate the need for more detailed information and how the logistics of such a network might work.

This author believes that with proper exploration and implementation, the past experience alternative can become a viable seismic qualification procedure.

Design Team Judgment

An economic saving can also be realized as a result of the design team's judgment. The design team has the option of qualifying equipment through

FIGURE 3.5. Tabletop sterilizer requiring design team judgment for seismic qualification.

inspection, adequate architectural detailing, or suggested installation techniques for which the facility owner will be responsible. Figure 3.5 shows an example sterilizer. The design team can make suggestions to the facility owner with respect to installation details for this type of equipment. Modular installation details for this type of equipment are discussed in Chapter 4. Design team judgment is also needed when off-the-shelf commodity items or daily use items, such as electric light switches, telephones, filing cabinets, and vending machines are used. These items can often be qualified by design team judgment and simple installation recommendations to the facility owners. Design team judgment is an excellent method that can be used to review existing facilities that have not received adequate seismic detailing.

Combined Qualification Approach

In some cases it may be advantageous to combine certain portions of the qualification methods discussed earlier. For example, a combined qualification program might be for an emergency power supply system. The normal operating loads on the diesel engine itself are generally greater than the seismic environment. Therefore, the engine might be qualified by static analysis for base anchorage, assuming that vibration isolation is not used. The radiator portion of the driving engine is generally a cantilevered structure on the front of the skid and may be dynamically sensitive. This characteristic would justify a dynamic analysis of the radiator to determine its natural frequency, and possible excursions and to determine methods of bracing it.

The power transfer panel often contains dynamically sensitive switching that does not lend itself easily to mathematical modeling, which may justify some kind of seismic testing. After these items have all been qualified, the design team may apply their judgment in installing flexible fuel lines.

The abbreviated example presented here shows how the design team might arrive at the justification for combining the approaches for a completely qualified system or individual piece of equipment. This approach can often save both time and money for the overall qualification requirements established for a building program.

Backfitting Essential Facilities

A quick tour through most existing facilities sadly illustrates a lack of seismic awareness for the building equipment. One is especially disturbed on examining essential facilities.

The A.I.A. Research Corporation postulated in a 1977 study that essential facilities such as metropolitan hospitals, police stations, and fire stations can expect an increase in the request for services on the order of between 300 and 700 percent as a result of a significant earthquake. Most existing essential facilities are not ready for such an event. The newer facilities are getting better as building codes begin to address equipment requirements. Most essential facilities were, however, built prior to the initiation of equipment requirements.

It is not at all uncommon to see emergency lighting loosely sitting atop filing cabinets, vibration isolated emergency power supplies without motion restraints, and piping without lateral bracing. The list of black marks goes on and on. Unfortunately, the chance for the survival of many existing essential facilities does not appear to be very good.

There is only one major code that has pointedly addressed the problem of backfitting. As discussed in Appendix 1, existing elevators in California all require modifications to their seismic resistance. The reader is referred to Appendix 1 for further discussion on the particular pros and cons of the elevator code.

If we truly expect our existing essential facilities to remain operational, campaigns for altering the code requirements must be sought. This backfitting approach is fairly common practice in the nuclear power industry and should be employed at least for critical equipment within essential facilities. As a beginning, the programs could be fairly simple. Architects and engineers versed in seismic qualification can make tours of facilities documenting the deficiencies that are most likely to place a facility out of commission in the event of an earthquake. The price to pay for an operating hospital is rather small when compared with the potential harm that may result from one that is incapable of operating.

Facility administrators may apply many of the diagrammatic installation details of Chapter 4 for effective backfit programs. Facility personnel can, themselves in many cases, spot deficiencies and correct the situation relatively easily and inexpensively. Such a program might include installation of shelf parapets on casework to restrain shelved items and to anchor filing cabinets to prevent their toppling. While an architect or engineer may be useful in pointing out and prescribing intricate details, the normal operating staff can provide quite a number of minor changes that will significantly increase the facility's potential for operability.

SEISMIC DESIGN CATEGORIES

Seismic design categories allow the design team to rationally specify seismic design procedures for various types of equipment. We learn in subsequent sections how to apply the seismic category concept so that essential facilities will remain operational during and after a major earthquake. Simply writing a general code provision such as the UBC 1979 does not give us the necessary assurance that the facilities will remain operational. To attain the needed assurance, a seismic plan must be developed. The plan should be consistent so that all members of the design team will be conversing with the same vocabulary. This section outlines a basic method by which the design team in conjunction with the owner can identify and seismically categorize the nonstructural elements in an essential facility.

First, the architect, together with the other members of the design team, and the facility owners review all the facility functions from a systems point of view. This involves defining which functions are critical to the facility operation, which are only required for smoothness of operation, and which fall into the miscellaneous category. The functional review leads to the identification of the various operational systems, which are comprised of the individual equipment items. The following is an example list of the types of systems under consideration:

- Air handling.
- Ceiling (lighting, acoustics, etc.).
- Communication.
- Data processing.
- Emergency power supply.
- Fire protection.
- Heating supply.
- Kitchen.
- Medical.
- Office.

- Piping.
- Records retrieval.
- Security monitoring.
- Sewage disposal.
- Various life support.
- Vertical circulation (elevators).
- Water supply.

This list of systems is incomplete and continues on for many different types of facilities. Each of the systems listed above is composed of individual elements. Emergency power supply systems (unless D.C. powered), for instance, generally have a driving engine, a generator, vibration isolation (always in conjunction with motion restraints), a starter mechanism, battery charging unit, exhaust provisions, distribution network, marked convenience outlets, fuel supply, and fuel storage capabilities. Some organized method is required to assess the relative importance of the various systems and items necessary for the overall operation of the essential facility. To meet this requirement, the seismic categories given in Table 3.2 are proposed. To identify the system and equipment, a dual letter designation is suggested. Therefore, from Table 3.2 we know that the previous example of the emergency power supply system can be placed in seismic category "A." The battery charger that is a component of that system, however, would be best classed in seismic category "C." Putting the two designations together in the dual letter format yields a seismic category for both the system and the equipment of "A-C," as shown in Figure 3.6. The letter on the left desig-

TABLE 3.2. Seismic Category Definitions

Seismic Category	Definition
Critical equipment "A"	Systems or equipment that are required for the operation of the essential facility, life support, or where failure will directly and adversely affect the function of other critical systems or equipment
Support equipment "B"	Systems or equipment required for support functions; the facility can operate on a limited basis if a failure occurs
Support equipment "C"	Systems or equipment required for prolonged operation of the facility on a day-to-day basis
Support equipment "D"	All portable systems or equipment not in seismic category "A"
Miscellaneous equipment "E"	Convenience or miscellaneous systems or equipment

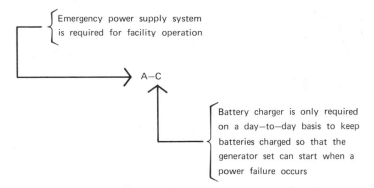

FIGURE 3.6.   System and equipment seismic category identification.

nates the seismic category of the overall system, while the letter on the right designates the seismic category of the individual piece of equipment. Figure 3.7 diagrammatically shows the decision process required for the use of the seismic category method.

Once the equipment and its operational system have been seismically categorized, the design team is ready to apply their resources to qualify the equipment. The seismic qualification approach depends on several factors:

- System category.
- Equipment category.
- Essential functions that equipment must perform.
- Expected earthquake intensity.
- Expected level of shaking at equipment or subcomponent location.
- Inherent design of equipment.
- Life expectancy of the facility.
- Mounting characteristics of equipment.
- Past experience of equipment.
- Proximity of equipment to other equipment items.

When these items are taken into account, the design team can rationally define the seismic category for both the system and the individual piece of equipment.

The systems and equipment section of this chapter suggests seismic categories for approximately 150 systems and equipment items. The list may seem long, but in reality is rather short. The seismic category provides for a common ground for the qualification of many different types of equipment.

The seismic categories are then used for establishing a basis for a logical seismic qualification approach and are applied to the seismic design specifications discussed in the next section.

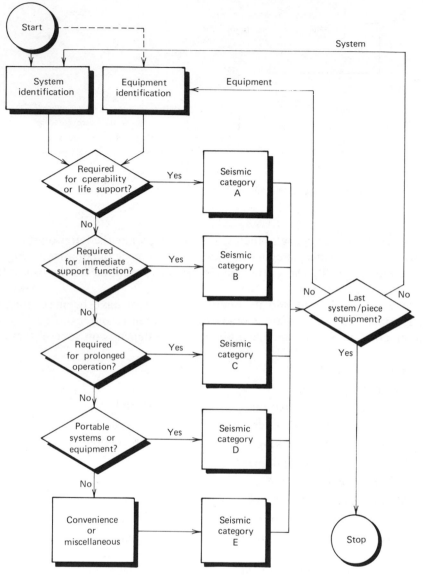

FIGURE 3.7.   Seismic category decision process.

SEISMIC DESIGN SPECIFICATION MODELS

Once the design team has successfully determined the seismic categories for the facility operational systems and the individual equipment items that comprise its systems, they need a specific design approach to qualify or plan the facility. In this section, two seismic design specifications are proposed for use by design teams: Seismic Design Specification-1 (SDS-1) and Seismic Design Specification-2 (SDS-2). SDS-1 applies to systems and equipment that can be included in seismic category "A" or "B," while SDS-2 applies to systems and equipment that can be included in seismic category "C," "D," or "E."

Floor response spectra, equipment location response spectra and static design coefficients must be supplied to the individual or organization performing the seismic qualification by the design team. This applies to both SDS-1 and SDS-2. The type of information required depends mostly on whether the qualification required will be performed at the system level, equipment level, or subcomponent level. The author wishes to express his gratitude to Wyle Laboratories for their assistance in the development of these model specifications.

Seismic Design Specification-1 (SDS-1)

SDS-1 provides the seismic design criteria for equipment included in seismic categories "A" and "B." For consistency within this specification, the following terms are defined:

- **Design Acceleration** (A)   The equipment acceleration value expressed in units of gravity.
- **Design Spectrum**   The curves of maximum responses of equipment subjected to a specific earthquake (critical equipment and support equipment). The response spectrum format is generally expressed as acceleration, velocity, or displacement versus frequency (1/period) for a designated equipment damping.
- **Design Team**   Architects/engineers responsible for the essential facility design.
- **Equipment Weight** (W)   The total equipment weight including contents such as oil and water.
- **Critical Equipment Earthquake**   The maximum ground motion possible at the site from any earthquake. Consideration must be given for expected effects felt at the site due to the distance from the hypocenter, local geology, soil conditions, and so forth.
- **Support Equipment Earthquake**   The maximum ground motion likely to occur within the life expectancy of the facility.
- **Owner**   Individual, organization, company, or government agency responsible for managing the facility.
- **Seismic Category**
  A.   Systems or equipment required for the operation of the facility and

life support, or whose failure could directly and adversely affect the function of other required systems or equipment.

   B.   Systems or equipment that are required for support functions. The facility can operate on a limited basis if a failure occurs.

- **Seismic Force (*F*)**   Static force coefficient that represents the equivalent seismic inertial force.
- **Seismic Load**   The inertial force applied to equipment at its location as a result of an earthquake and the building/component interface.
- **Seismic Qualification**   Demonstrated means by which the equipment can be shown to resist the expected seismic loads in a manner that satisfies the design objectives.

Depending on the seismic category and the nature of the equipment, seismic qualification can be demonstrated by the following means:

- Shake table tests.
- Mathematical analysis.
  - Dynamic.
  - Equivalent static coefficient.
- Past experience.
- Design team judgment.
- Any combination of the above.

The critical equipment earthquake design response spectrum will be used where seismic qualification is required of equipment that must remain operational because of its importance to the functioning of the facility after an earthquake or where it involves life support functions. Seismic qualification of other equipment in seismic category "A" and "B" will use the support equipment earthquake design response spectrum. The design team supplies those performing the seismic qualification (the manufacturer, owner, private consultant, or the design team itself) with the following requirements and information:

- Applicable critical equipment earthquake, required response spectra (with damping).
- Applicable support equipment earthquake, required response spectra (with damping).
- Applicable floor design accelerations.
- Applicable equivalent static coefficients.
- Identification of which critical damping factors are to be used with respect to the design spectra.

The organization or individual bidding on the seismic qualification shall supply the design team with the following information prior to commencing the seismic qualification program:

- Price quotation for the proposed method of qualification.
- Detailed qualification plan.
  - If by analysis: method of analysis, design criteria for analysis, and professional credentials of those conducting the analysis.
  - If by test: detailed test plan, mounting of specimen, operational loads of specimen, proposed instrumentation of specimen, test machine capabilities, choice of waveform(s).
  - If by past experience: detailed characteristics of the experience (test or analysis report).
  - If by design team judgment: detailed explanation of rationale behind the decision.

The design team and owners shall review and approve all proposals submitted prior to the actual seismic qualification program. The owner shall notify the successful bidder when the seismic qualification program is to begin for each equipment item. The individual or organization performing the seismic qualification shall notify the owner prior to any seismic test programs so that the design team or owner may witness the test at their option.

The seismic qualification program is to be carried out as mutually agreed upon by the bidder (his proposal plus required revisions where applicable), the design team, and the owner. A seismic qualification procedure flow loop is shown in Figure 3.8. The qualification report shall include proof of the following:

- Seismic test (where applicable).
  - Mounting to simulate actual in-service installation.
  - Test in two perpendicular horizontal axes and vertical axis. The test may be conducted in each direction independently. The design spectra must be multiplied by 1.5 to compensate for the single axis test. If the test is conducted in two axes simultaneously, the design spectra must be multiplied by 1.2 for life support equipment or any equipment required for the continued operation of life support equipment, and two support equipment earthquake (SEE) complete tests must be conducted prior to conducting the critical equipment earthquake (CEE). The lower level SEE is more likely to occur over the lifetime of the facility and their cumulative effects should be examined to ensure operability after the higher level CEE.

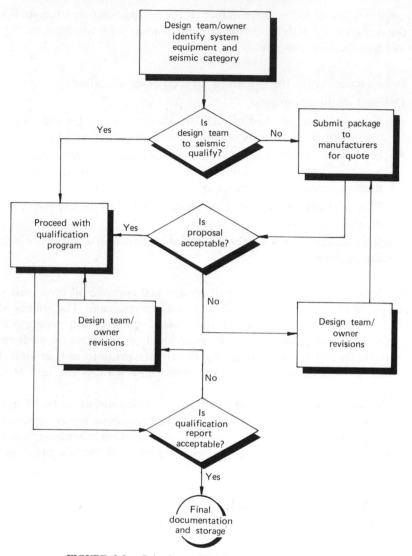

FIGURE 3.8. Seismic qualification procedure flow loop.

- Waveforms
  1. Resonant frequency search (sine sweep test) at not greater than 2 octaves/minute from 0.1 to 33 hertz in all three major axes. This test is to be a low-amplitude test and should be conducted in all test programs.
  2. Earthquake tests shall employ waveforms that are appropriate to the specific test. Seismic tests may employ any of the following:

time histories, random motions, random motion with superimposed sine beats, or complex waveform tests. In any case, the required response spectrum must be enveloped by the test response spectrum for the test to be considered valid.

- Criteria for testing.
- Test machine calibration.
- Installation details.
- Continuous monitoring of equipment before, during, and after seismic test to assure operation of equipment.
- Performance anomalies observed.
- Equipment that has been earthquake tested is acceptable for installation provided that the test does not result in nonrepairable failure.

● Mathematical analysis (where applicable).
- Seismic forces are imposed in two horizontal axes and the vertical axis.
- Installation details must be provided.
- Static ($F = AW$) or dynamic analyses may be performed.
- Operational loads must be combined with seismic loads.
- Total seismic response from three axis analysis may be taken as square root sum of the squares of each individual response to each direction.
- Maximum deflection of the equipment must be reported.

All equipment is to be designed and installed to resist overturning, sliding, and content spillage. Friction due to elastomeric feet and so forth will be ignored in all seismic calculations and tests. The results of the seismic qualification required by SDS-1 must be documented in a step-by-step form suitable for audit.

Seismic Design Specification-2 (SDS-2)

SDS-2 provides the seismic design criteria for equipment included in seismic categories "C," "D," and "E." The seismic categories are defined below:

C. Systems or equipment required for prolonged operation of the facility on a day-to-day basis.
D. All portable systems or equipment not in seismic category "A."
E. Convenience or miscellaneous systems or equipment.

The principal requirement of SDS-2 is that the equipment remain anchored during and after an earthquake. Anchorage of the equipment and subcomponents, which include items such as base plates, equipment enclosures, cabinets, legs, supporting structures, connections, rolling stock, and cantilevered supports, will be designed to resist sliding, rolling, and content spillage and/or overturning. There will not be an allowance made for friction at the base of equipment supports because of the use of elastomeric feet to resist sliding.

The equipment shall be analyzed in two mutually perpendicular horizontal directions and the vertical direction. A static coefficient analysis ($F = AW$) is the recommended approach. The design acceleration coefficient "A" is to be provided by the design team. The total seismic response from the three axis analysis may be taken as the square root sum of the squares of the individual responses to each direction.

A proposal is not required of the bidder for SDS-2. The bidder, however, must submit a seismic qualification price quotation to the design team prior to execution of the analysis. The design team and owner will review, approve, and notify the successful bidder when to perform the analysis.

Several other methods are available by which the equipment can be qualified when the SDS-2 recommended approach is not applicable:

- Seismic testing.
- Dynamic mathematical analysis.
- Past experience.
- Design team judgment.
- Any combination of the above.

Those performing the qualification must coordinate any approach other than that recommended with the design team.

SYSTEMS AND EQUIPMENT QUALIFICATION

This section discusses the systems and equipment with which this book is concerned. The major systems are divided alphabetically and are listed below:

- Access floor systems.
- Air handling systems.
- Communication systems.
- Data processing systems.
- Elevator systems.
- Emergency power supply systems.
- Fire protection systems.

- Kitchen systems.
- Lighting systems.
- Medical systems.
- Piping systems.
- Suspended ceiling systems.
- Water systems.
- Miscellaneous equipment.

The discussion of each system contains a short general statement and a suggestion for the overall system seismic category. Examples of the types of facilities most likely to contain major components of the system in question are also listed.

Individual examples of equipment within each system are included in alphabetical order. A general statement describes important aspects of the equipment item, and photographic examples are given for most pieces of equipment. Suggestions for the equipment seismic category, the appropriate seismic specification and seismic qualification approaches are given for each equipment item. It should be noted that these suggestions should be examined in detail by the design team for each application to ensure adequate seismic qualification. Reference is also made in most cases to Chapter 4 for diagrammatic examples of installation details. In some cases, reference is made to "similar generic" details where specific details are not supplied.

Subjective scenarios are given for the degree of damage possible and the type or consequence of damage for inadequately protected equipment. The relationships given here are in part based on observations of past performance of specific pieces of equipment in earthquakes and in part on extrapolations from the performance of similar pieces of equipment. None of the scenarios is directed toward individual manufacturers nor their equipment.

For some equipment items reference is made to photographic examples of damaged equipment that can be found in the final pages of Chapter 3. The author wishes to apologize for the quality of some of the photographs, as the originals were either unavailable or untraceable, which necessitated copies being made directly from previous publications.

This section of Chapter 3 and all of Chapter 4 are designed to be used in handbook fashion by design professionals, governing agencies, facility owners, manufacturers, and students. First the applicable system is identified and then the individual equipment within that system is located. The design team then notes the suggestions for the qualification approach listed for the equipment. Reference is easily made by figure number to Chapter 4 for the suggested diagrammatic installation details. This handbook approach should lead to new and existing (backfitted) facilities that are much more likely to remain operational after earthquakes.

Access Floor Systems

Access or raised floor systems generally imply lift-out floor modules. Freestanding, as well as earthquake resistant, interlocking systems are available and are commonly used where data processing or communications equipment is required. If the access floor system should fail during an earthquake, the equipment that it supports is also more than likely to fail.

SYSTEM SEISMIC CATEGORY

- "A" critical system.

SYSTEM FOUND IN

- Business establishments.
- Communication centers.
- Computing/data processing centers.
- Emergency operating centers.
- Fire stations.
- Government administration buildings.
- Hospitals.
- Police stations.

Access Floor Systems

Floor Panels

Access floor panels support heavy, expensive pieces of equipment and must remain in place so that the supported equipment will not be damaged.

EQUIPMENT SEISMIC CATEGORY

- "A" critical equipment.

SEISMIC SPECIFICATION

- SDS-1.

SEISMIC QUALIFICATION APPROACH

- Equivalent static coefficient analysis.
- Design team judgment.
  - Select a floor manufacturer that has built-in earthquake protection measures.

REFERENCE FIGURES FOR INSTALLATION DETAILS

- 4.1, 4.2.

RELATIVE DEGREE OF DAMAGE OF INADEQUATELY PROTECTED EQUIPMENT

- Moderate to major.

MOST LIKELY TYPE OR CONSEQUENCE OF DAMAGE FOR
INADEQUATELY PROTECTED EQUIPMENT

- Dislodged panels.
- Dislodged equipment due to shifting floor.
- Floor collapse.
- Inoperable equipment supported by floor.
- General cleanup required.

Access Floor Systems

Stanchions

Floor stanchions (Figure 3.9) are the support columns for the access floor system. For the best performance, the stanchions should be anchored to the subfloor, braced between each other, and anchored to the floor panels.

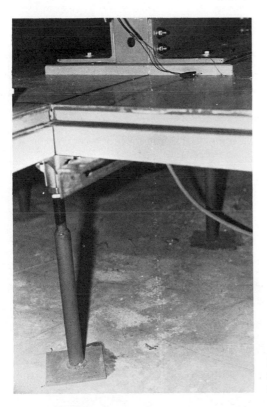

FIGURE 3.9. Access floor stanchion that is anchored neither to the subfloor nor to the access floor frame.

EQUIPMENT SEISMIC CATEGORY

- "A" critical equipment.

SEISMIC SPECIFICATION

- SDS-1.

SEISMIC QUALIFICATION APPROACH

- Equivalent static coefficient analysis.
- Design team judgment.
  - Select a floor manufacturer that has built-in earthquake protection measures.

REFERENCE FIGURES FOR INSTALLATION DETAILS

- 4.1, 4.2.

RELATIVE DEGREE OF DAMAGE OF INADEQUATELY PROTECTED EQUIPMENT

- Minor to moderate.

MOST LIKELY TYPE OR CONSEQUENCE OF DAMAGE FOR
INADEQUATELY PROTECTED EQUIPMENT

- Shifting of floor.
- Collapse of floor.
- Inoperable equipment supported by floor.
- General cleanup required.

Air Handling Systems

Air handling equipment is found in most building types. Most facilities can operate on a limited basis if failures should occur.

SYSTEM SEISMIC CATEGORY

- "B" support system.

SYSTEM FOUND IN

- All facilities.

Air Handling Systems

Air Grilles, Registers, and Diffusers

These items (Figure 3.10) are not required for the successful operation of the system, but do present a danger to personnel if they fall.

FIGURE 3.10.   Air grille that does not have a safety wire suspending it from the structure above. Photo courtesy Ruhnau ·Evans·Ruhnau·Associates.

EQUIPMENT SEISMIC CATEGORY

- "E" miscellaneous equipment.

SEISMIC SPECIFICATION

- SDS-2.

SEISMIC QUALIFICATION APPROACH

- Design team judgment.
  - Screw and nut installation recommended; sheet metal screws and friction fits common fail.
  - Retaining safety wire.

RELATIVE DEGREE OF DAMAGE OF INADEQUATELY PROTECTED EQUIPMENT

- Minor.

MOST LIKELY TYPE OR CONSEQUENCE OF DAMAGE FOR
INADEQUATELY PROTECTED EQUIPMENT

- Fallen registers.
- Potential for personnel injury.
- General cleanup.

REFERENCE FIGURES FOR EXAMPLE OF DAMAGED EQUIPMENT

- 3.148, 3.152, 3.153.

Air Handling Systems

Chillers and Heaters

Chillers (Figure 3.11) and heaters are generally complicated equipment items that must be at least anchored to prevent fluid leaks. Flexible inlet and outlet connections improve their survivability.

FIGURE 3.11.   Physical plant air conditioning unit with flexible connections. The air conditioning unit as well as its subcomponents should be considered.

EQUIPMENT SEISMIC CATEGORY

- "B" support equipment.

SEISMIC SPECIFICATION

- SDS-1.

SEISMIC QUALIFICATION APPROACH

- Equivalent static coefficient analysis.
  - Base anchorage.
- Dynamic analysis.
  - Manufacturer generic qualification.

REFERENCE FIGURES FOR INSTALLATION DETAILS

- 4.5, 4.6.

RELATIVE DEGREE OF DAMAGE OF INADEQUATELY PROTECTED EQUIPMENT

- Moderate to major.

MOST LIKELY TYPE OR CONSEQUENCE OF DAMAGE FOR
INADEQUATELY PROTECTED EQUIPMENT

- Shifted equipment.
- Toppled equipment.
- Broken supply lines.
- Inoperative equipment.
- Flooding potential with some equipment.

REFERENCE FIGURE FOR EXAMPLE OF DAMAGED EQUIPMENT

- 3.151.

Air Handling Systems

Duct Work

Duct work (Figure 3.12) needs lateral bracing and flexible joints at junctions. Bracing should not bend or wrap around other equipment.

EQUIPMENT SEISMIC CATEGORY

- "C" support equipment.

SEISMIC SPECIFICATION

- SDS-2.

SEISMIC QUALIFICATION APPROACH

- Design team judgment.
  - Flexible connections at rigid building interfaces.
- Equivalent static coefficient analysis.
  - Support and lateral bracing.

REFERENCE FIGURES FOR INSTALLATION DETAILS

- 4.3, 4.4, 4.96 and Appendix 3.

RELATIVE DEGREE OF DAMAGE OF INADEQUATELY PROTECTED EQUIPMENT

- Moderate.

MOST LIKELY TYPE OR CONSEQUENCE OF DAMAGE FOR INADEQUATELY PROTECTED EQUIPMENT

- Duct work collapse.
- Ruptured duct work if flexible connections are not provided at rigid building interfaces.

FIGURE 3.12. Ductwork showing transverse bracing.

- Inoperative portions of air system.
- General cleanup required.

REFERENCE FIGURES FOR EXAMPLE OF DAMAGED EQUIPMENT

- 3.149, 3.150.

Air Handling Systems

Fan Units, Floor-Mounted

Air handling fan units (Figure 3.13) range from small to very large. If vibration isolators are employed, motion restraints are required. Consideration should also be given to line connections.

EQUIPMENT SEISMIC CATEGORY

- ''B'' support equipment.

SEISMIC SPECIFICATION

- SDS-1.

SEISMIC QUALIFICATION APPROACH

- Equivalent static coefficient analysis.
  - Base anchorage.
- Dynamic analysis.
  - Employ motion restraints if vibration isolation is used.

FIGURE 3.13. Roof-mounted fan unit showing fixed mounting and flexible service line connections.

REFERENCE FIGURES FOR INSTALLATION DETAILS

● 4.5, 4.78, 4.79, 4.80, 4.81, 4.82, 4.83, 4.84, 4.85, 4.86.

RELATIVE DEGREE OF DAMAGE OF INADEQUATELY PROTECTED EQUIPMENT

● Minor to moderate.

MOST LIKELY TYPE OR CONSEQUENCE OF DAMAGE FOR
INADEQUATELY PROTECTED EQUIPMENT

● Shifting or overturning of the fan unit.
● Resonance of equipment if vibration isolators are used without motion restraints (may result in major damage).
● Possibly inoperative fan units.
● Cleanup required.

REFERENCE FIGURES FOR EXAMPLE OF DAMAGED EQUIPMENT

● 3.151, 3.174, 3.177.

Air Handling Systems

Fan Units, Suspended

Suspended fans (Figure 3.14) obviously present special problems should they fall. The fan unit should be anchored to a structural wall wherever possible.

EQUIPMENT SEISMIC CATEGORY

● "B" support equipment.

SEISMIC SPECIFICATION

● SDS-1.

SEISMIC QUALIFICATION APPROACH

● Equivalent static coefficient analysis.
  • If rigidly mounted (natural frequency generally above 30 hertz).
  • Lateral bracing.
● Dynamic analysis.
  • If flexibly mounted.
  • Motion restraints are required if vibration isolation is employed.

REFERENCE FIGURES FOR INSTALLATION DETAILS

● 4.6, 4.78, 4.79, 4.80, 4.81, 4.82, 4.83, 4.84, 4.85.

FIGURE 3.14. Suspended fan unit with motion restraints on the vibration isolators, flexible duct connections, and fixed anchorage to the adjacent structural wall at the bottom left.

RELATIVE DEGREE OF DAMAGE OF INADEQUATELY PROTECTED EQUIPMENT

● Major.

MOST LIKELY TYPE OR CONSEQUENCE OF DAMAGE FOR
INADEQUATELY PROTECTED EQUIPMENT

● Damage potential to adjacent equipment from excessive swaying.
● Fallen fan units.
● Potential for injured personnel.
● Potential for equipment damaged by falling fan units.
● General cleanup required.

Air Handling Systems

Mixing Boxes

Mixing boxes (Figure 3.15) need lateral bracing and flexible joints at ducting interface.

EQUIPMENT SEISMIC CATEGORY

● "C" support equipment.

SEISMIC SPECIFICATION

● SDS-2.

SEISMIC QUALIFICATION APPROACH

● Design team judgment.
   • Flexible connections.
● Equivalent static coefficient analysis.
   • Bracing and support.

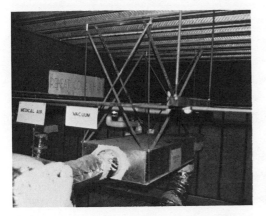

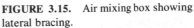

FIGURE 3.15.   Air mixing box showing lateral bracing.

REFERENCE FIGURES FOR INSTALLATION DETAILS

● 4.7.

RELATIVE DEGREE OF DAMAGE OF INADEQUATELY PROTECTED EQUIPMENT

● Minor.

MOST LIKELY TYPE OR CONSEQUENCE OF DAMAGE FOR
INADEQUATELY PROTECTED EQUIPMENT

● Dislodged mixing box.
● Severed ducting connections.
● Inoperative portions of air system.
● General cleanup required.

Air Handling Systems

Vibration Isolators

Vibration isolation on reciprocating air handling equipment (Figure 3.16) should receive motion restraint considerations. A commonly held misconception is that vibration isolation allows equipment to "float" through the earthquake unscathed. Nothing could be further from the truth! Vibration isolators commonly have spring systems with natural frequencies coincident with the vibrations produced by earthquakes. Without some snubbing device, the spring-mass system is likely to go into resonance and eventually destroy itself through wild and uncontrolled vibration. Also see Motion Restraint Systems in this chapter.

EQUIPMENT SEISMIC CATEGORY

● "B" support equipment.

FIGURE 3.16. Air handling unit illustrating vibration isolation on the left and motion restraint (snubber) on the right.

SEISMIC SPECIFICATION

- SDS-1.

SEISMIC QUALIFICATION APPROACH

- Dynamic analysis.
  - For vibration isolation, if not provided by vibration isolation manufacturer. Employ motion restraint of some sort.
- Follow vibration isolation manufacturer's suggestions for motion restraint.

REFERENCE FIGURES FOR INSTALLATION DETAILS

- 4.78, 4.79, 4.80, 4.81, 4.82, 4.83, 4.84, 4.85.

RELATIVE DEGREE OF DAMAGE OF INADEQUATELY PROTECTED EQUIPMENT

- Major.

MOST LIKELY TYPE OR CONSEQUENCE OF DAMAGE FOR
INADEQUATELY PROTECTED EQUIPMENT

- Equipment that employs vibration isolators without motion restraint almost always fails during strong motion.
- Inoperative equipment likely.
- Serious damage may result from "flying" springs that have failed.
- General cleanup required.

REFERENCE FIGURES FOR EXAMPLES OF DAMAGED EQUIPMENT

- 3.174, 3.175, 3.177.

Communications Systems

Adequate communications are a necessary function following any major disaster. Communications systems require thorough qualification programs to provide reasonable assurance that they will remain operational. In cases such as ambulance and fire control, trucks must rely on police facilities to direct them to the emergency area by way of unblocked roads. Without operational communication systems, calls for help might go unanswered.

SYSTEM SEISMIC CATEGORY

- "A" critical system.

SYSTEM FOUND IN

- Communication centers.
- Emergency operating centers.
- Fire stations.
- Government administration buildings.
- Hospitals.
- Police stations.

Communications Systems

Antennas

Antenna whip can lead to its potential collapse and is a major concern for this type of equipment (see Figure 3.17).

EQUIPMENT SEISMIC CATEGORY

- "A" critical equipment.

SEISMIC SPECIFICATION

- SDS-1.

SEISMIC QUALIFICATION APPROACH

- Dynamic analysis.
  - Base anchorage determination.
  - Guying where possible.

REFERENCE FIGURE FOR INSTALLATION DETAILS

- 4.8.

RELATIVE DEGREE OF DAMAGE OF INADEQUATELY PROTECTED EQUIPMENT

- Moderate.

FIGURE 3.17.   Roof top antenna without midpoint bracing.

MOST LIKELY TYPE OR CONSEQUENCE OF DAMAGE FOR
INADEQUATELY PROTECTED EQUIPMENT

- Severe whipping of antenna.
- Collapse of antenna.
- Inoperative equipment.

Communications Systems

Antennas, Cavitated

Cavitated antennas (Figure 3.18) generally fit within freestanding cabinets or are attached directly to walls. They are relatively lightweight.

EQUIPMENT SEISMIC CATEGORY

- ''A'' critical equipment.

SEISMIC SPECIFICATION

- SDS-1.

SEISMIC QUALIFICATION APPROACH

- Equivalent static coefficient analysis.
    - Anchorage within cabinet or to building walls.

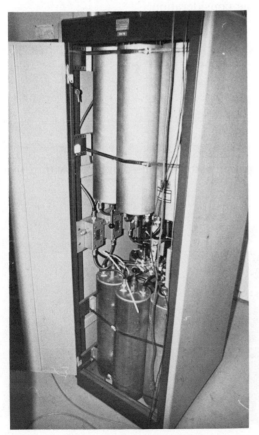

FIGURE 3.18. These lightweight cavitated antennae are not adequately anchored within their cabinet and the cabinet has not been base anchored. Note the base anchorage holes provided by the cabinet manufacturer.

REFERENCE FIGURE FOR INSTALLATION DETAILS

- 4.14.

RELATIVE DEGREE OF DAMAGE OF INADEQUATELY PROTECTED EQUIPMENT

- Minor.

MOST LIKELY TYPE OR CONSEQUENCE OF DAMAGE FOR
INADEQUATELY PROTECTED EQUIPMENT

- Dislodged antenna.
- Inoperable equipment.
- General cleanup.

Communications Systems

Cabinets, Freestanding

Free-standing cabinets (Figure 3.19) can commonly house critical equipment and should receive adequate seismic consideration.

EQUIPMENT SEISMIC CATEGORY

- Varies—"A," "B," "C," or "E."

SEISMIC SPECIFICATION

- SDS-1.

SEISMIC QUALIFICATION APPROACH

- Seismic test.
  - If dynamically sensitive critical equipment is contained within the cabinet.

FIGURE 3.19. Freestanding cabinets require base anchorage and top bracing wherever possible. These are unsecured examples.

- Equivalent static coefficient analysis.
  - For cabinet anchorage. Anchor at bottom as a minimum precaution and, if possible, at top also.

REFERENCE FIGURES FOR INSTALLATION DETAILS

- 4.9, 4.14, 4.15.

RELATIVE DEGREE OF DAMAGE OF INADEQUATELY PROTECTED EQUIPMENT

- Minor to major.

**MOST LIKELY TYPE OR CONSEQUENCE OF DAMAGE FOR
INADEQUATELY PROTECTED EQUIPMENT**

- Fallen or dislodged cabinets.
- Damaged equipment contained within the cabinet.
- Possibly inoperative equipment.
- General cleanup required.

Communications Systems

Cable Trays

Cable trays (Figure 3.20) keep a myriad of cable out of the work space and can be used in lieu of, or in conjunction with, access floors.

EQUIPMENT SEISMIC CATEGORY

- "B" support equipment.

SEISMIC SPECIFICATION

- SDS-1.

SEISMIC QUALIFICATION APPROACH

- Equivalent static coefficient analysis.

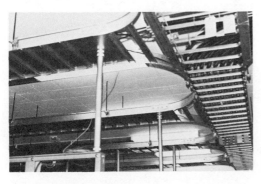

FIGURE 3.20. These cable trays lack lateral bracing.

- Cable tray support system.
- Design team judgment.
  - Cable trays must pass freely through partition walls.
  - Slack in cables must be specified.
  - Adequate vertical support.
  - Lateral bracing of tray.

REFERENCE FIGURES FOR INSTALLATION DETAILS

- 4.9, 4.10, 4.11.

RELATIVE DEGREE OF DAMAGE OF INADEQUATELY PROTECTED EQUIPMENT

- Minor to moderate.

MOST LIKELY TYPE OR CONSEQUENCE OF DAMAGE FOR
INADEQUATELY PROTECTED EQUIPMENT

- Collapsed tray systems.
- Ripped wiring.
  - Fire potential if lines are hot.
- Potential for inoperative equipment.
- General cleanup required.

Communications Systems

Call Recorders, Floor-Mounted

Floor-mounted call recorders (Figure 3.21) should be base anchored and well restrained to prevent pounding of adjacent systems.

EQUIPMENT SEISMIC CATEGORY

- "B" support equipment.

SEISMIC SPECIFICATION

- SDS-1.

SEISMIC QUALIFICATION APPROACH

- Dynamic analysis.
  - Of cabinet structure.
- Equivalent static coefficient analysis.
  - Base anchorage.
  - Cabinet sway restraint.
  - Top anchorage if feasible.

FIGURE 3.21. This call recorder is base anchored; top bracing, however, has not been provided and the unit is likely to bang against the wall to the left during an earthquake.

REFERENCE FIGURES FOR INSTALLATION DETAILS

- 4.9, 4.14, 4.15, 4.35.

RELATIVE DEGREE OF DAMAGE OF INADEQUATELY PROTECTED EQUIPMENT

- Minor to moderate.

MOST LIKELY TYPE OR CONSEQUENCE OF DAMAGE FOR INADEQUATELY PROTECTED EQUIPMENT

- Toppled recorder.
  - Potentially inoperative.
- Damage due to pounding of adjacent items.
- General cleanup required.

Communications Systems

Call Recorders, Table-Mounted

Table-mounted (Figure 3.22) call recorders should be securely fixed to the table, not a writing desk, and the table itself should be securely anchored.

FIGURE 3.22. This tabletop call recorder is anchored to its base, which is in turn anchored to the floor.

EQUIPMENT SEISMIC CATEGORY

● "B" support equipment.

SEISMIC SPECIFICATION

● SDS-1.

SEISMIC QUALIFICATION APPROACH

● Equivalent static coefficient analysis.
● Design team judgment.
  • Designers and operators commonly allow tall and slender equipment such as this to be mounted, unsecured, to a writing desk. This invites sure inoperability of the equipment as a result of toppling.

REFERENCE FIGURES FOR INSTALLATION DETAILS

● 4.21, 4.61, 4.62, 4.63.

RELATIVE DEGREE OF DAMAGE OF INADEQUATELY PROTECTED EQUIPMENT

● Minor to moderate.

MOST LIKELY TYPE OR CONSEQUENCE OF DAMAGE FOR
INADEQUATELY PROTECTED EQUIPMENT

● Toppled equipment.
  • Potentially inoperable.

- Toppled table.
- General cleanup required.

Communications Systems

Consoles

The console (Figure 3.23) is the major operator work station. Mechanical and electrical equipment within the cabinet may require additional earthquake qualification even if the frame itself is adequately anchored. Anchorage to an interlocked access floor or the subfloor must be considered.

EQUIPMENT SEISMIC CATEGORY

- "A" critical equipment.

SEISMIC SPECIFICATION

- SDS-1.

SEISMIC QUALIFICATION APPROACH

- Seismic test.
  - For operability.
- Dynamic analysis.
  - Mainframe integrity.
- Equivalent static coefficient analysis.
  - Equipment anchorage.
- Design team judgment.
  - Doors and drawers require positive latches.

REFERENCE FIGURES FOR INSTALLATION DETAILS

- 4.16, 4.17, 4.58.

FIGURE 3.23. The equipment main-frame as well as the internal subcompo-nents require earthquake consideration to prevent failure. This console is not base anchored.

RELATIVE DEGREE OF DAMAGE OF INADEQUATELY PROTECTED EQUIPMENT

- Moderate to major.

MOST LIKELY TYPE OR CONSEQUENCE OF DAMAGE FOR
INADEQUATELY PROTECTED EQUIPMENT

- Dislodged console, ripped wiring.
- Doors and drawers opened.
- Inoperable internal components.
- General cleanup required.

Communications Systems

Direct Current Power Packs

Batteries or D.C. power packs are commonly required in communication facilities for specific pieces of equipment as a backup power supply (other than the standby generator).

EQUIPMENT SEISMIC CATEGORY

- "A" critical equipment.

SEISMIC SPECIFICATION

- SDS-1.

SEISMIC QUALIFICATION APPROACH

- Equivalent static coefficient analysis.
  - Battery rack anchorage.
- Design team judgment.
  - Since batteries are heavy, provide adequate anchorage to the building structure not just to a partition wall as shown in Figure 3.24.
  - Specify slack in all cable.
  - Specify cushioning material, such as Styrofoam, between batteries and the rack.

REFERENCE FIGURES FOR INSTALLATION DETAILS

- 4.30, 4.31.

RELATIVE DEGREE OF DAMAGE OF INADEQUATELY PROTECTED EQUIPMENT

- Moderate to major.

MOST LIKELY TYPE OR CONSEQUENCE OF DAMAGE FOR
INADEQUATELY PROTECTED EQUIPMENT

- Toppled battery racks.
- Broken cables.

FIGURE 3.24. These battery racks have been anchored to a nonstructural wall and have not been base anchored. Note the base anchorage provisions provided by the manufacturer. The electric cables have also been installed taut rather than with slack.

- Broken battery cases.
- Inoperative equipment, acid spills.
- General cleanup.

Communication Systems

Dispatch

Dispatch operations must begin immediately following a major disaster. All dispatch related equipment (Figure 3.25) must be considered.

EQUIPMENT SEISMIC CATEGORY

- "A" critical equipment.

SEISMIC SPECIFICATION

- SDS-1.

SEISMIC QUALIFICATION APPROACH

- Seismic test.

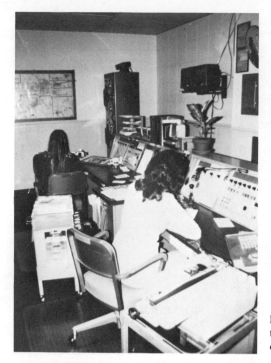

FIGURE 3.25. Dispatch units are vitally important and should be carefully qualified to protect against failure.

- For dynamically sensitive equipment that must demonstrate operability.
- Dynamic analysis.
  - For equipment racks, console frames, and so on.
- Equivalent static coefficient analysis.
  - For equipment anchorage.
- Design team judgment.
  - For desk top equipment, and so on.

REFERENCE FIGURES FOR INSTALLATION DETAILS

- 4.16, 4.17, 4.18, 4.19, 4.20, 4.21, 4.58, 4.61, 4.62, 4.63.

RELATIVE DEGREE OF DAMAGE OF INADEQUATELY PROTECTED EQUIPMENT

- Moderate to major.

MOST LIKELY TYPE OR CONSEQUENCE OF DAMAGE FOR INADEQUATELY PROTECTED EQUIPMENT

- Inoperative dynamically sensitive equipment.
- Dislodged equipment.
- Toppled equipment.

- Potentially inoperative equipment.
- General cleanup required.

Communications Systems

Equipment Racks

Equipment racks are typically mathematically modeled as inverted pendulums and are commonly anchored only at their base. This anchorage system leads to wild vibration patterns during the earthquake and increases the equipment failure potential.

EQUIPMENT SEISMIC CATEGORY

- "A" critical equipment.

SEISMIC SPECIFICATION

- SDS-1.

SEISMIC QUALIFICATION APPROACH

- Seismic test.

FIGURE 3.26. Consideration must be given to both communication racks and the dynamically sensitive equipment that they hold.

- • For dynamically sensitive subcomponents.
- Dynamic analysis.
  - • For frame integrity.
- Equivalent static coefficient analysis.
  - • For base and top anchorage.
  - • Top bracing where possible.
- Design team judgment.
  - • Use through bolts with nuts and lock washers and locate heavy, sensitive equipment toward the bottom of the rack.

REFERENCE FIGURES FOR INSTALLATION DETAILS

- 4.14, 4.15.

RELATIVE DEGREE OF DAMAGE OF INADEQUATELY PROTECTED EQUIPMENT

- Minor to major.

MOST LIKELY TYPE OR CONSEQUENCE OF DAMAGE FOR
INADEQUATELY PROTECTED EQUIPMENT

- Deformed racks.
- Toppled racks.
- Racks can deflect enough to collide with other equipment.
- Failure of dynamically sensitive subcomponents.
- System inoperability.
- General cleanup.

Communications Systems

INTERAC Operations

Volunteer communications operations within emergency operating centers are often neglected in the design phase (see Figure 3.27). They do, however, represent an important communication function that deserves consideration.

FIGURE 3.27. Leaving communication equipment loose atop unanchored tables as in this example will likely result in inoperative equipment.

EQUIPMENT SEISMIC CATEGORY

- "B" support equipment.

SEISMIC SPECIFICATION

- SDS-1.

SEISMIC QUALIFICATION APPROACH

- Design team judgment.
  - Anchor tables.
  - Restrain table top equipment.
  - Restrain adjacent equipment such as that shown in Figure 3.27.

REFERENCE FIGURES FOR INSTALLATION DETAILS

- 4.12, 4.13, 4.18, 4.19, 4.20, 4.21, 4.58, 4.61, 4.62, 4.63.

RELATIVE DEGREE OF DAMAGE OF INADEQUATELY PROTECTED EQUIPMENT

- Minor to moderate.

MOST LIKELY TYPE OR CONSEQUENCE OF DAMAGE FOR
INADEQUATELY PROTECTED EQUIPMENT

- Overturned tables.
- Equipment falling off table top.
- Inoperative equipment.
- General cleanup required.

Communications Systems

Storage, Ad Hoc

Unprotected ad hoc storage (Figure 3.28) adjacent to any expensive/
sensitive equipment has the potential of damaging otherwise well protected
equipment.

EQUIPMENT SEISMIC CATEGORY

- "E" miscellaneous equipment.

SEISMIC SPECIFICATION

- SDS-2.

SEISMIC QUALIFICATION APPROACH

- Equivalent static coefficient analysis.
  - Shelf case anchorage.
- Design team judgment.

FIGURE 3.28. Ad hoc storage such as that shown here can easily topple and damage adjacent critical equipment.

- Do not store equipment where it can damage other equipment if at all possible.
- Use shelf restrainers to keep items on their shelves.

REFERENCE FIGURES FOR INSTALLATION DETAILS

- 4.53, 4.54, 4.55, 4.56, 4.73, 4.74, 4.76, 4.102, 4.103, 4.104, 4.108, 4.109.

RELATIVE DEGREE OF DAMAGE OF INADEQUATELY PROTECTED EQUIPMENT

- Minor to major.

MOST LIKELY TYPE OR CONSEQUENCE OF DAMAGE FOR
INADEQUATELY PROTECTED EQUIPMENT

- Toppled shelves.
- Spilled shelved items.
- Potentially inoperable equipment.
- General cleanup required.

REFERENCE FIGURE FOR EXAMPLE OF DAMAGED EQUIPMENT

- 3.198.

Communications Systems

Teletypes

Hard copy communication equipment (Figure 3.29) such as teletypes generally requires base anchorage of the equipment frame and top bracing to prevent pounding against walls.

EQUIPMENT SEISMIC CATEGORY

● "B" support equipment.

SEISMIC SPECIFICATION

● SDS-1.

SEISMIC QUALIFICATION APPROACH

● Equivalent static coefficient analysis.
  • Base anchorage.
● Design team judgment.
  • Mount equipment far enough away from the adjacent wall to keep the equipment from pounding against it during strong motion or provide top bracing.

REFERENCE FIGURES FOR INSTALLATION DETAILS

● 4.16, 4.17, 4.58.

RELATIVE DEGREE OF DAMAGE OF INADEQUATELY PROTECTED EQUIPMENT

● Minor to moderate.

FIGURE 3.29. Teletype and other telecommunications equipment should always receive aseismic consideration. This example has not received any.

MOST LIKELY TYPE OR CONSEQUENCE OF DAMAGE FOR
INADEQUATELY PROTECTED EQUIPMENT

- Toppling.
  - Potentially inoperative equipment.
- General cleanup required.

Communications Systems

Test Equipment, Portable

Portable test equipment (Figure 3.30) presents a special hazard to stationary equipment (critical or support) and personnel if left unrestrained when not in use.

EQUIPMENT SEISMIC CATEGORY

- "D" support equipment.

SEISMIC SPECIFICATION

- SDS-2.

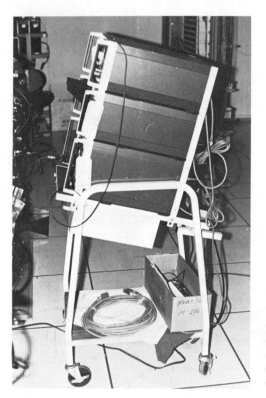

FIGURE 3.30. Unattended portable test equipment can easily collide with adjacent critical equipment leaving it inoperable. Wheel locks do not prevent earthquake damage.

SEISMIC QUALIFICATION APPROACH

- Design team judgment.
  - Provide for fixed protected storage space.
- Equivalent static coefficient analysis.
  - For restraining mechanism when not in use.

REFERENCE FIGURES FOR INSTALLATION DETAILS

- 4.12, 4.13, 4.61, 4.62, 4.63, 4.68, 4.69.

RELATIVE DEGREE OF DAMAGE OF INADEQUATELY PROTECTED EQUIPMENT

- Minor to major.

MOST LIKELY TYPE OR CONSEQUENCE OF DAMAGE FOR
INADEQUATELY PROTECTED EQUIPMENT

- Toppled test equipment.
- Damage to fixed equipment from collision.
- Potential personnel injury.
- Inoperable equipment.
- General cleanup required.

REFERENCE FIGURE FOR EXAMPLE OF DAMAGED EQUIPMENT
- 3.170.

Communications Systems

Wiring

Wiring and conduit should always be left with slack (Figure 3.31) to allow for differential movement.

EQUIPMENT SEISMIC CATEGORY

- "A" critical equipment.

SEISMIC SPECIFICATION

- SDS-1.

SEISMIC QUALIFICATION APPROACH

- Design team judgment.
  - Specify slack in all wiring and conduit, especially at all building and equipment interfaces.

REFERENCE FIGURES FOR INSTALLATION DETAILS

- 4.9, 4.34.

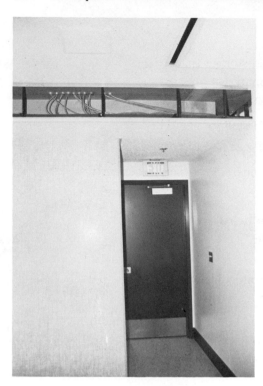

FIGURE 3.31. All wiring should be installed with slack. Shown here is wiring with proper slack, as well as wiring that has been pulled tight.

RELATIVE DEGREE OF DAMAGE OF INADEQUATELY PROTECTED EQUIPMENT

● Moderate to major.

MOST LIKELY TYPE OR CONSEQUENCE OF DAMAGE FOR
INADEQUATELY PROTECTED EQUIPMENT

● Ripped wiring.
 · Fire potential.
● Potentially inoperative equipment.

Data Processing Systems

Data processing systems are generally not needed for the continued operating capabilities of most facilities. The equipment is, however, quite expensive, difficult to repair, and time consuming to replace. Failure of seemingly minor equipment such as cooling systems can render the entire data processing center useless. For these reasons, it is necessary to adhere closely to qualification procedures.

SYSTEM SEISMIC CATEGORY

● "B" support system.

SYSTEM FOUND IN

● Business establishments.
● Computing/data processing centers.
● Government administration buildings.
● Schools.

Data Processing Systems

Cooling Equipment

Cooling capabilities are an absolute must for all computers. Cooling can be accomplished with refrigerated air or integral water systems (see Figure 3.32).

EQUIPMENT SEISMIC CATEGORY

● "A" critical equipment.

FIGURE 3.32. This computer cooling unit has been installed without base anchorage, which effectively jeopardizes the entire data processing system if it should fail.

SEISMIC SPECIFICATION

- SDS-1.

SEISMIC QUALIFICATION APPROACH

- Equivalent static coefficient analysis.
  - Base anchorage of all pumps, cooling towers, and so on.
- Design team judgment.
  - Leave enough slack in all water lines to allow for movement.
  - Use flexible hose where possible or flexible connectors.

REFERENCE FIGURES FOR INSTALLATION DETAILS

- 4.9, 4.16, 4.17, 4.58, 4.88, 4.92, 4.93.

RELATIVE DEGREE OF DAMAGE OF INADEQUATELY PROTECTED EQUIPMENT

- Minor to major.

MOST LIKELY TYPE OR CONSEQUENCE OF DAMAGE FOR
INADEQUATELY PROTECTED EQUIPMENT

- Inoperable equipment.
- Flooding if water supply lines should sever as a result of being installed without slack or flexible connections.
- General cleanup required.

Data Processing Systems

Disc Storage and Compilers

Disc storage (Figure 3.33) and compilers generally have wide footings and are thus likely to be fairly stable. They do, however, require base anchorage to prevent sliding.

EQUIPMENT SEISMIC CATEGORY

- "B" support equipment.

SEISMIC SPECIFICATION

- SDS-1.

SEISMIC QUALIFICATION APPROACH

- Equivalent static coefficient analysis.
  - Base anchorage of cabinet to access floor or subfloor.
- Seismic test/dynamic analysis.
  - Manufacturer may wish to undertake a more ambitious generic qualification program to assure equipment operability and frame/component integrity.

FIGURE 3.33. These disc storage units have not received protection against sliding or overturning.

REFERENCE FIGURES FOR INSTALLATION DETAILS

- 4.9, 4.14, 4.15, 4.34, 4.58.

RELATIVE DEGREE OF DAMAGE OF INADEQUATELY PROTECTED EQUIPMENT

- Minor to moderate.

MOST LIKELY TYPE OR CONSEQUENCE OF DAMAGE FOR INADEQUATELY PROTECTED EQUIPMENT

- Sliding cabinet.
- Possibly inoperable equipment due to torn wires.
- General cleanup required.

Data Processing Systems

Interface, Input/Output

Input/output periferals such as keyboards, line printers, and CRTs (Figure 3.34) need to be restrained from sliding and toppling.

EQUIPMENT SEISMIC CATEGORY

- "B" support equipment.

FIGURE 3.34. Input/output devices such as the CRT terminal shown here are typically left sitting atop tables without any earthquake protection.

SEISMIC SPECIFICATION

● SDS-1.

SEISMIC QUALIFICATION APPROACH

● Equivalent static coefficient analysis.
   • Anchorage or adequate restraint of tables, and so forth.

REFERENCE FIGURES FOR INSTALLATION DETAILS

● 4.9, 4.14, 4.15, 4.34, 4.58, 4.61, 4.62, 4.63, 4.110.

RELATIVE DEGREE OF DAMAGE OF INADEQUATELY PROTECTED EQUIPMENT

● Minor to moderate.

MOST LIKELY TYPE OR CONSEQUENCE OF DAMAGE FOR
INADEQUATELY PROTECTED EQUIPMENT

● Dislodged equipment.
● Toppled equipment.
● Possibly inoperable equipment.
● General cleanup required.

Data Processing Systems

Tape Drives

Tape drives (Figure 3.35) are generally tall and slender pieces of equipment susceptible to toppling.

EQUIPMENT SEISMIC CATEGORY

● ''B'' support equipment.

SEISMIC SPECIFICATION

● SDS-1.

SEISMIC QUALIFICATION APPROACH

● Equivalent static coefficient analysis.
   • Base anchorage.
   • Top bracing where possible.
● Seismic test/dynamic analysis.
   • Manufacturer may wish to undertake a more ambitious generic qualification to assure equipment operability and frame/component integrity.

REFERENCE FIGURES FOR INSTALLATION DETAILS

● 4.9, 4.14, 4.15, 4.34, 4.35.

FIGURE 3.35. Tape drives are likely to topple during an earthquake if they are not properly anchored.

RELATIVE DEGREE OF DAMAGE OF INADEQUATELY PROTECTED EQUIPMENT

● Minor to moderate.

MOST LIKELY TYPE OR CONSEQUENCE OF DAMAGE FOR INADEQUATELY PROTECTED EQUIPMENT

● Toppled equipment.
● Possibly inoperable equipment.
● General cleanup required.

Data Processing Systems

Tape Storage

Tape storage units (Figure 3.36) should be base anchored and shelf restrainers should be employed to keep the tapes on the shelves.

EQUIPMENT SEISMIC CATEGORY

● "C" support equipment.

SEISMIC SPECIFICATION

● SDS-2.

FIGURE 3.36. The contents of tape storage units such as these may spill if shelf restrainers are not employed along with base anchorage and top bracing.

SEISMIC QUALIFICATION APPROACH

- Equivalent static coefficient analysis.
  - Base anchorage.
  - Top bracing and anchorage.
  - Longitudinal "×" bracing.
- Design team judgment.
  - Provide restraints to keep tapes on the shelves.

REFERENCE FIGURES FOR INSTALLATION DETAILS

- 4.35, 4.53, 4.54, 4.55, 4.56, 4.73, 4.104.

RELATIVE DEGREE OF DAMAGE OF INADEQUATELY PROTECTED EQUIPMENT

- Minor.

MOST LIKELY TYPE OR CONSEQUENCE OF DAMAGE FOR
INADEQUATELY PROTECTED EQUIPMENT

- Toppled shelves.
- Tapes spilled onto floors.
- General cleanup required.

Elevator Systems

Passenger elevators are divided into two general types: the oil-hydraulic elevators and the traction elevators. The former generally performs well with a minimum of earthquake protection. The traction type elevators, however, are more complicated and consequently are more likely to fail as a result of strong shaking. Most of the emphasis of legislation currently in existence for elevators is directed at the traction elevators (see Appendix 1, CAC 24-7 for California retrofit requirements). Elevator manufacturers may wish to consider generic seismic test programs to qualify the elevator system so that it will remain operational after a major earthquake. Current California legislation will have to be revised so that elevators are not restricted from operation (see Appendix 1). Oil-hydraulic elevators are commonly composed of the following earthquake sensitive subcomponents:

- Pump.
- Electric motor.
- Electric control panel.
- Hydraulic control.
- Cab guide rails.

Traction elevators are commonly composed of the following earthquake sensitive subcomponents:

- Hoist machine.
- Motor control panel.
- Motor generator.
- Selector panel.
- Counterweights and rails.
- Car and car rails.

SYSTEM SEISMIC CATEGORY

- "A" critical system.

SYSTEM FOUND IN

- All multistory facilities.

Elevator Systems—Oil-Hydraulic Elevator

Cab Guide Rails

Inadequately designed and anchored guide rails can fail and cause elevator system inoperability (see Figure 3.37).

EQUIPMENT SEISMIC CATEGORY

- "A" critical equipment.

FIGURE 3.37. Cab guide rails need adequate anchorage provisions to prevent failure. Shown here is an inadequate example.

SEISMIC SPECIFICATION

- SDS-1.

SEISMIC QUALIFICATION APPROACH

- Equivalent static coefficient analysis.
  - Rail anchorage.
  - Rail characteristics.

REFERENCE FIGURE FOR INSTALLATION DETAILS

- 4.22.

RELATIVE DEGREE OF DAMAGE OF INADEQUATELY PROTECTED EQUIPMENT

- Minor to moderate.

MOST LIKELY TYPE OR CONSEQUENCE OF DAMAGE FOR
INADEQUATELY PROTECTED EQUIPMENT

- Derailed car.
- Inoperable elevator.

Elevator Systems—Oil-Hydraulic Elevators

Electric Control Panel

If the electric control panel (Figure 3.38) is not solid-state and if it contains dynamically sensitive subcomponents such as relays or mercury switches, it should be qualified to prove operability.

EQUIPMENT SEISMIC CATEGORY

- "A" critical equipment.

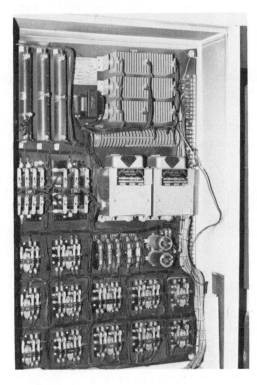

FIGURE 3.38. Electric control panels are securely attached to walls. Some of the subcomponents may be dynamically sensitive.

SEISMIC SPECIFICATION

● SDS-1.

SEISMIC QUALIFICATION APPROACH

● Equipment static coefficient analysis.
  • Base anchorage and top bracing if solid state.
● Seismic test.
  • To prove operability if dynamically sensitive subcomponents are used.

REFERENCE FIGURE FOR INSTALLATION DETAILS

● 4.25.

RELATIVE DEGREE OF DAMAGE OF INADEQUATELY PROTECTED EQUIPMENT

● Minor to moderate.

MOST LIKELY TYPE OR CONSEQUENCE OF DAMAGE FOR INADEQUATELY PROTECTED EQUIPMENT

● Shifted equipment.

- Toppled equipment.
- Inoperable as a result of failure of dynamically sensitive subcomponents (transient failure).

Elevator Systems—Oil-Hydraulic Elevators

Electric Motor

The electric motor (Figure 3.39) is a subcomponent of the hydraulic control unit and must remain in place for the elevator system to be operational.

EQUIPMENT SEISMIC CATEGORY

- "A" critical equipment.

SEISMIC SPECIFICATION

- SDS-1.

SEISMIC QUALIFICATION APPROACH

- Equivalent static coefficient analysis.
  - Base anchorage.

REFERENCE FIGURE FOR INSTALLATION DETAILS

- 4.28.

RELATIVE DEGREE OF DAMAGE OF INADEQUATELY PROTECTED EQUIPMENT

- Minor to moderate.

MOST LIKELY TYPE OR CONSEQUENCE OF DAMAGE FOR
INADEQUATELY PROTECTED EQUIPMENT

- Shifted motor.
- Inoperable elevator.

FIGURE 3.39. The electric motor should be firmly anchored to the control frame as shown here.

Elevator Systems—Oil-Hydraulic Elevators

Hydraulic Control Unit

The hydraulic control unit (Figure 3.40) must remain in place for the elevator system to be operational and hydraulic supply lines must maintain low flange loads to prevent oil spills.

EQUIPMENT SEISMIC CATEGORY

- "A" critical equipment.

SEISMIC SPECIFICATION

- SDS-1.

SEISMIC QUALIFICATION APPROACH

- Equivalent static coefficient analysis.
  - Base anchorage.
- Design team judgment.
  - Flexible fluid supply lines.

REFERENCE FIGURE FOR INSTALLATION DETAILS

- 4.28.

RELATIVE DEGREE OF DAMAGE OF INADEQUATELY PROTECTED EQUIPMENT

- Minor to moderate.

MOST LIKELY TYPE OR CONSEQUENCE OF DAMAGE FOR
INADEQUATELY PROTECTED EQUIPMENT

- Shifted equipment.
- Broken supply lines.
- Inoperable elevator.

FIGURE 3.40. The hydraulic control unit must be anchored and flexible line connections should be employed. Hard connections, as shown here, are likely to fail.

REFERENCE FIGURE FOR EXAMPLE OF DAMAGED EQUIPMENT

- 3.162.

Elevator Systems—Oil-Hydraulic Elevators

Pump

The hydraulic pump (Figure 3.41) is a subcomponent of the hydraulic control unit and must remain in place to be operational, and fluid supply lines must maintain low flange loads to prevent oil spills.

EQUIPMENT SEISMIC CATEGORY

- "A" critical equipment.

SEISMIC SPECIFICATION

- SDS-1.

SEISMIC QUALIFICATION APPROACH

- Equivalent static coefficient analysis.
  - Fixed anchorage.
- Dynamic analysis.
  - Vibration isolation (motion restraint required).
- Design team judgment.
  - Flexible fluid supply lines.

REFERENCE FIGURE FOR INSTALLATION DETAILS

- 4.28.

RELATIVE DEGREE OF DAMAGE OF INADEQUATELY PROTECTED EQUIPMENT

- Minor to moderate.

FIGURE 3.41. The hydraulic pump, and other subcomponents, should be securely anchored to the control frame.

MOST LIKELY TYPE OR CONSEQUENCE OF DAMAGE FOR
INADEQUATELY PROTECTED EQUIPMENT

- Shifted equipment.
- Broken supply lines.
- Inoperable elevator.

Elevator Systems—Traction Elevators

Car Rails

Elevator car rails (Figure 3.42) have generally performed better in past earthquakes than the counterweights. Ayres and Sun (1973) only reported 18 cars out of their rails or out of alignment as a result of the 1971 San Fernando earthquake. This observation does not mean that car rails should not be considered; they should.

EQUIPMENT SEISMIC CATEGORY

- "A" critical equipment.

SEISMIC SPECIFICATION

- SDS-1.

SEISMIC QUALIFICATION APPROACH

- Equivalent static coefficient analysis.
  - Rail anchorage.
- Stress analysis.
  - For the rail.
- Design team judgment.
  - Select guide shoe for shoes with earthquake provisions.

REFERENCE FIGURE FOR INSTALLATION DETAILS

- 4.22.

FIGURE 3.42. Car rail with roller guide. Photograph taken from car top.

- Minor to moderate.

MOST LIKELY TYPE OR CONSEQUENCE OF DAMAGE FOR
INADEQUATELY PROTECTED EQUIPMENT

- Derailed elevator cars.
- Inoperable elevators.

REFERENCE FIGURE FOR EXAMPLE OF DAMAGED EQUIPMENT

- 3.157.

Elevator Systems—Traction Elevators

Counterweight Guide Rails

Ayres and Sun (1973) reported 674 counterweights that had been dislodged from their guide rails as a result of the 1971 San Fernando earthquake. One-sixth of the counterweights actually damaged the elevator cars.

EQUIPMENT SEISMIC CATEGORY

- "A" critical equipment.

SEISMIC SPECIFICATION

- SDS-1.

SEISMIC QUALIFICATION APPROACH

- Equivalent static coefficient analysis.
  - Rail anchorage.
- Stress analysis.
  - On the rail.
- Design team judgment.
  - Specify proper counterweight roller guides (Figure 3.43) that have been designed for the earthquake environment.

REFERENCE FIGURES FOR INSTALLATION DETAILS

- 4.22, 4.23.

RELATIVE DEGREE OF DAMAGE OF INADEQUATELY PROTECTED EQUIPMENT

- Moderate to major.

MOST LIKELY TYPE OR CONSEQUENCE OF DAMAGE FOR
INADEQUATELY PROTECTED EQUIPMENT

- Potential for personnel injury.
- Counterweights derailed.

FIGURE 3.43. Counterweight guide rails showing rail anchorage and counterweight roller guides at top of counterweight.

- Counterweights damage elevator car.
- Inoperable elevator system.

REFERENCE FIGURES FOR EXAMPLES OF DAMAGED EQUIPMENT

- 3.154, 3.155, 3.156, 3.157, 3.158.

Elevator Systems—Traction Elevators

Hoist Machine

Hoist machine failure can cause the hoist cables to become entangled, which in turn makes the entire elevator system inoperable.

EQUIPMENT SEISMIC CATEGORY

- ''A'' critical equipment.

SEISMIC SPECIFICATION

- SDS-1.

SEISMIC QUALIFICATION APPROACH

- Equivalent static coefficient analysis.
    - Base anchorage of hoist machine to structural member (Figure 3.44).

REFERENCE FIGURE FOR INSTALLATION DETAILS

- 4.27.

RELATIVE DEGREE OF DAMAGE OF INADEQUATELY PROTECTED EQUIPMENT

- Minor to major.

MOST LIKELY TYPE OR CONSEQUENCE OF DAMAGE FOR
INADEQUATELY PROTECTED EQUIPMENT

- Shifted hoist machine.

FIGURE 3.44.   Gearless hoist machine showing base anchorage and cable guide.

- Toppled hoist machine.
- Tangled hoist cables.
- Inoperable elevator.

Elevator Systems—Traction Elevators

Motor Control Panel

Motor control panels with dynamically sensitive control devices (Figure 3.45) require more stringent qualification programs than panels with all solid-state subcomponents.

EQUIPMENT SEISMIC CATEGORY

- ''A'' critical equipment.

SEISMIC SPECIFICATION

- SDS-1.

SEISMIC QUALIFICATION APPROACH

- Equivalent static coefficient analysis.
  - Base anchorage of control panels with solid-state subcomponents.
- Seismic test.
  - Control panels with dynamically sensitive switches, and so on.

REFERENCE FIGURE FOR INSTALLATION DETAILS

- 4.25.

RELATIVE DEGREE OF DAMAGE OF INADEQUATELY PROTECTED EQUIPMENT

- Minor to major.

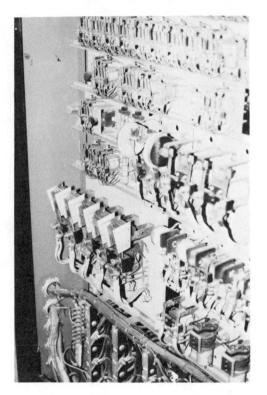

FIGURE 3.45. Motor control panels showing dynamically sensitive switches. This tall, slender panel is neither base anchored nor top braced.

MOST LIKELY TYPE OR CONSEQUENCE OF DAMAGE FOR INADEQUATELY PROTECTED EQUIPMENT

- Shifted equipment.
- Toppled equipment.
- False signaling.
- Inoperable equipment.

REFERENCE FIGURE FOR EXAMPLE OF DAMAGED EQUIPMENT

- 3.159.

Elevator Systems—Traction Elevators

Motor Generator

Motor generators have commonly failed in past earthquakes because of inadequate anchorage (see Figure 3.46). They must be base anchored and provided with flexible electric connections.

FIGURE 3.46. Unsecured motor generator sitting atop floor grating. The electric wiring to this generator has been installed without slack, which increases its failure potential.

EQUIPMENT SEISMIC CATEGORY

- "A" critical equipment.

SEISMIC SPECIFICATION

- SDS-1.

SEISMIC QUALIFICATION APPROACH

- Equivalent static coefficient analysis.

REFERENCE FIGURE FOR INSTALLATION DETAILS

- 4.29.

RELATIVE DEGREE OF DAMAGE OF INADEQUATELY PROTECTED EQUIPMENT

- Minor to major.

MOST LIKELY TYPE OR CONSEQUENCE OF DAMAGE FOR
INADEQUATELY PROTECTED EQUIPMENT

- Shifted equipment.
- Inoperable elevators.

REFERENCE FIGURES FOR EXAMPLES OF DAMAGED EQUIPMENT

- 3.159, 3.160, 3.161.

Elevator Systems—Traction Elevators

Selector Panel

Selector panels with dynamically sensitive subcomponents can cause the inoperability of the elevator system through false signaling, and so on if not adequately braced (Figure 3.47).

FIGURE 3.47. This selector panel has only been anchored to the floor grating, not the steel structure below. This is not adequate seismic protection.

EQUIPMENT SEISMIC CATEGORY

● "A" critical equipment.

SEISMIC SPECIFICATION

● SDS-1.

SEISMIC QUALIFICATION APPROACH

● Equivalent static coefficient analysis.
  · Base anchorage and top bracing for solid-state panels.
● Seismic test.
  · Panels with dynamically sensitive subcomponents.

REFERENCE FIGURE FOR INSTALLATION DETAILS

● 4.26.

RELATIVE DEGREE OF DAMAGE OF INADEQUATELY PROTECTED EQUIPMENT

● Minor to major.

- Shifted equipment.
- Toppled equipment.
- Pounding of adjacent equipment or walls if not properly braced.
- Inoperability due to false signaling, and so on (transient failure).

Emergency Power Supply Systems

Emergency power supply systems are the backbone of all facilities following most major disasters. Widespread power failures commonly accompany destructive earthquakes. More than any of the other systems discussed in this book, emergency power supplies need to be approached from the systems point of view for seismic qualification. A single failure of many subcomponents in the system can place the entire system out of operation. This can lead to the inoperability of other equipment items that may be required to perform critical functions. The interrelationships of functioning systems was demonstrated in the 1979 Imperial Valley earthquake. At one essential facility, a water main burst and flooded the emergency power supply room. As a result of the flooding, the emergency power supply system could not function and critical communications were hampered because of a lack of power. Inadequate qualification of the emergency power supply system and its components or adjacent systems will almost certainly lead to critical failures in future earthquakes.

SYSTEM SEISMIC CATEGORY

- "A" critical system.

SYSTEM FOUND IN

- Business establishments.
- Communication centers.
- Computing/data processing centers.
- Emergency operating centers.
- Fire stations.
- Government administration buildings.
- Hospitals.
- Police stations.
- Schools.

Emergency Power Supply Systems

Battery

To start larger systems, more than one battery is required. Battery racks should be adequately anchored and the batteries should be anchored or restrained within their racks (Figure 3.48). Electric cables should have plenty of slack and not be pulled tight for aesthetic reasons.

EQUIPMENT SEISMIC CATEGORY

- "A" critical equipment.

SEISMIC SPECIFICATION

- SDS-1.

SEISMIC QUALIFICATION APPROACH

- Equivalent static coefficient analysis.
  - For battery rack anchorage.
- Design team judgment.
  - For cable slack and battery restraint within the rack.

REFERENCE FIGURES FOR INSTALLATION DETAILS

- 4.30, 4.31.

RELATIVE DEGREE OF DAMAGE OF INADEQUATELY PROTECTED EQUIPMENT

- Minor to major.

MOST LIKELY TYPE OR CONSEQUENCE OF DAMAGE FOR
INADEQUATELY PROTECTED EQUIPMENT

- Battery racks may topple if not adequately anchored.

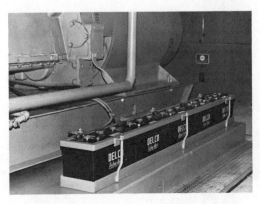

FIGURE 3.48. Batteries should be anchored as shown here and flexible electrical connections should be provided.

- Battery cells may crack if not restrained within their racks; acid may spill from cracked batteries.
- Electric cables tear loose from the terminals, often with battery case damage, if installed too tightly.
- Emergency power supply system cannot function.

Emergency Power Supply Systems

Exhaust Unit

Exhaust gases are harmful to facility personnel if leaks should develop. Flexible connections provide leak protection and should be provided between the manifold/muffler (Figure 3.49) and muffler/building interfaces.

EQUIPMENT SEISMIC CATEGORY

- "A" critical equipment.

SEISMIC SPECIFICATION

- SDS-1.

FIGURE 3.49. Flexible connections between the exhaust manifold and muffler should be employed as shown here. Lateral bracing also should have been provided for the muffler unit.

SEISMIC QUALIFICATION APPROACH

- Equivalent static coefficient analysis.
  - For suspension system.
  - Lateral bracing.
- Dynamic analysis.
  - To determine maximum displacements.
- Design team judgment.
  - Use manufacturer recommended flexible connections.

REFERENCE FIGURE FOR INSTALLATION DETAILS

- 4.42.

RELATIVE DEGREE OF DAMAGE OF INADEQUATELY PROTECTED EQUIPMENT

- Minor.

MOST LIKELY TYPE OR CONSEQUENCE OF DAMAGE FOR
INADEQUATELY PROTECTED EQUIPMENT

- Inadequate suspension system can cause silencer collapse.
- Rigid exhaust pipe connections are likely to fail.
- Potential for personnel injury if exhaust gases leak into the building.
- System is likely to remain operational even if silencer supports fail.

Emergency Power Supply Systems

Fuel Supply Equipment

The day tank (Figure 3.50) generally holds enough fuel for 24 hours of continued operation. To assure that variations in the time frame requirement are met it is necessary to refer to local codes. The day tank may be buried or fixed to the walls or floor of the facility. Damage would generally not be expected if the day tank is securely anchored.

EQUIPMENT SEISMIC CATEGORY

- "A" critical equipment.

SEISMIC SPECIFICATION

- SDS-1.

SEISMIC QUALIFICATION APPROACH

- Equivalent static coefficient analysis.
  - Fixed day tank.
- No specific requirements.
  - Buried day tank.

FIGURE 3.50. Small day tank showing vibration isolation without motion restraints and with flexible fuel lines.

REFERENCE FIGURES FOR INSTALLATION DETAILS

- 4.32, 4.33.

RELATIVE DEGREE OF DAMAGE OF INADEQUATELY PROTECTED EQUIPMENT

- None to minor.

MOST LIKELY TYPE OR CONSEQUENCE OF DAMAGE FOR
INADEQUATELY PROTECTED EQUIPMENT

- If the day tank is dislodged, fuel may be spilled.
- Potential fire hazard from spilled fuel.
- If the day tank is dislodged, the emergency power supply may be left inoperative.

Emergency Power Supply Systems

Fuel Supply Lines

The major mode of failure for fuel supply lines is rupturing at rigid connections. Distribution metering boxes and fuel pumps should be fixed. All supply lines from the day tank to the emergency power generator should have

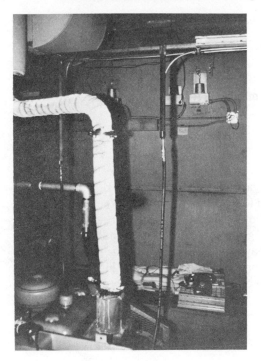

FIGURE 3.51. Flexible fuel lines should be used to assure generator operability. Rubber lines are used here.

flexible connections at every junction (Figure 3.51). Flexible connections (rubber, braided, or copper tubing) should have slack; that is, they should not be stretched tight for aesthetic reasons.

EQUIPMENT SEISMIC CATEGORY

- ''A'' critical equipment.

SEISMIC SPECIFICATION

- SDS-1.

SEISMIC QUALIFICATION APPROACH

- Design team judgment.
  - Provide flexible connections with plenty of slack.

REFERENCE FIGURE FOR INSTALLATION DETAILS

- 4.36.

RELATIVE DEGREE OF DAMAGE OF INADEQUATELY PROTECTED EQUIPMENT

- None if flexible connections are used.
- Moderate to major if rigid connections are used.

- Ruptured fuel line.
- Potential fire hazard from spilled fuel.
- The emergency power supply may be left inoperative if the fuel supply lines rupture.

Emergency Power Supply Systems

Generator Set

The generator set may be diesel, gasoline or turbine powered. The engine and generator are usually subjected to loads far in excess of the seismic loads during its normal operating conditions. They are usually attached to a skid and must be restrained either through fixed anchorage or snubbed vibration isolators. Moderate damage that may render the generator set inoperative is generally due to the failure of vibration isolation mountings that do not employ motion restraints (Figure 3.52).

EQUIPMENT SEISMIC CATEGORY

- "A" critical equipment.

SEISMIC SPECIFICATION

- SDS-1.

SEISMIC QUALIFICATION APPROACH

- Equivalent static coefficient analysis.
  - Fixed anchorage.
- Dynamic analysis.
  - Vibration isolation.
- Seismic test.
  - Manufacturers may consider generic programs for the entire system.

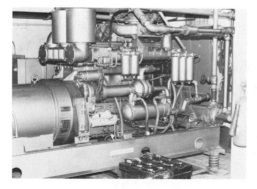

FIGURE 3.52. This emergency power supply has been installed without motion restraints for the vibration isolators. Note the batteries adjacent to the frame. They are likely to be damaged during an earthquake.

REFERENCE FIGURES FOR INSTALLATION DETAILS

- 4.37, 4.38, 4.39, 4.40, 4.41, 4.78, 4.79, 4.80, 4.81, 4.82, 4.83, 4.84, 4.85, 4.86.

RELATIVE DEGREE OF DAMAGE OF INADEQUATELY PROTECTED EQUIPMENT

- Minor to major.

MOST LIKELY TYPE OR CONSEQUENCE OF DAMAGE FOR
INADEQUATELY PROTECTED EQUIPMENT

- Vibration isolation failure.
- Battery failure.
- Fuel line failure.
- Other peripheral equipment failure.
- Equipment may be left inoperative as a result of the above mentioned failures.

Emergency Power Supply Systems

Generators, Portable

Portable generators should be stored securely when not in use to prevent damage (see Figure 3.53).

EQUIPMENT SEISMIC CATEGORY

- "A" critical equipment.

SEISMIC SPECIFICATION

- SDS-1.

FIGURE 3.53. Portable generators should not be haphazardly stored on wheel carts as shown here. They should be securely stored to prevent toppling, and so forth.

SEISMIC QUALIFICATION APPROACH

- Equivalent static coefficient analysis.
  - Of restrainers.
- Design team judgment.
  - Provide restraining system to keep this type of portable equipment stationary when not in use.

REFERENCE FIGURE FOR INSTALLATION DETAILS

- 4.69.

RELATIVE DEGREE OF DAMAGE OF INADEQUATELY PROTECTED EQUIPMENT

- Minor to major.

MOST LIKELY TYPE OR CONSEQUENCE OF DAMAGE FOR
INADEQUATELY PROTECTED EQUIPMENT

- Runaway equipment.
- Collision with other equipment.
- Toppled equipment.
- Potentially inoperative generator.
- General cleanup required.

Emergency Power Supply Systems

Power Transfer Panel

Power transfer from the generator set to the building distribution network goes through the power transfer panel. This unit commonly contains switches and relays that are sensitive to dynamic motions, which could cause operational failures (see Figure 3.54).

EQUIPMENT SEISMIC CATEGORY

- "A" critical equipment.

FIGURE 3.54. Electric transfer panels with dynamically sensitive subcomponents require sophisticated qualification procedures.

SEISMIC SPECIFICATION

- SDS-1.

SEISMIC QUALIFICATION APPROACH

- Seismic test
  - For both fixed and vibration isolation mounted panels. The latter is generally found in Civil Defense emergency operating centers only.
  - For panels with dynamically sensitive switches.
- Equivalent static coefficient analysis.
  - For all other panels.

REFERENCE FIGURES FOR INSTALLATION DETAILS

- 4.34, 4.35.

RELATIVE DEGREE OF DAMAGE OF INADEQUATELY PROTECTED EQUIPMENT

- Minor to moderate.

MOST LIKELY TYPE OR CONSEQUENCE OF DAMAGE FOR
INADEQUATELY PROTECTED EQUIPMENT

- Operational failure of dynamically sensitive switches.
- Structural failure of panel is possible if vibration isolation is used without motion restraint.

Emergency Power Supply Systems

Vibration Isolation

Vibration isolators keep uncomfortable vibrations from being transmitted to the building during normal operation of reciprocating equipment. These isolators are, however, extremely sensitive to most earthquake motions (see Figure 3.55). Also see Motion Restraint Systems in this chapter.

FIGURE 3.55. Employment of vibration isolation always requires accompanying motion restraint equipment. This spring system does not have motion restraints and has been installed immediately adjacent to the starting batteries, which are likely to be damaged during strong motion.

EQUIPMENT SEISMIC CATEGORY

- "A" critical equipment.

SEISMIC SPECIFICATION

- SDS-1.

SEISMIC QUALIFICATION APPROACH

- Dynamic analysis.
  - By manufacturer.
- Seismic test.
  - For dynamically sensitive equipment such as power transfer panel if vibration isolation units are used.

REFERENCE FIGURES FOR INSTALLATION DETAILS

- 4.39, 4.40, 4.41, 4.78, 4.79, 4.80, 4.81, 4.82, 4.83, 4.84, 4.85, 4.86.

RELATIVE DEGREE OF DAMAGE OF INADEQUATELY PROTECTED EQUIPMENT

- Major.

MOST LIKELY TYPE OR CONSEQUENCE OF DAMAGE FOR
INADEQUATELY PROTECTED EQUIPMENT

- Resonance often occurs with major damage resulting.
- Inoperative emergency power supply system.

REFERENCE FIGURE FOR EXAMPLE OF DAMAGED EQUIPMENT

- 3.173.

Fire Protection Systems

Fire protection systems are highly varied in their application and complexity. The following general types are available:

- Chemical suppression.
- Hand-held extinguishers.
- Smoke and gas detectors.
- Water distribution networks.

Earthquake protection of the various fire protection systems can clearly reduce potential earthquake losses.

SYSTEM SEISMIC CATEGORY

- "A" critical system.

SYSTEM FOUND IN

- Business establishments.
- Communication centers.
- Computing/data processing centers.
- Emergency operating centers.
- Fire stations.
- Government administration buildings.
- Hospitals.
- Police stations.
- Schools.

Fire Protection Systems

Chemical Fire Suppression Units

These units usually contain the following elements:

- Chemical extinguisher tank.
- Connection to distribution system.
- Distribution system.

The units are generally not susceptible to dynamic motions as long as they remain anchored and if flexible connections are provided at structural interfaces (Figure 3.56).

EQUIPMENT SEISMIC CATEGORY

- "A" critical equipment.

SEISMIC SPECIFICATION

- SDS-1.

SEISMIC QUALIFICATION APPROACH

- Equivalent static coefficient analysis.
  - For anchorage of chemical bottles.
- Design team judgment.
  - For anchorage of distribution system and flexible connections.

REFERENCE FIGURE FOR INSTALLATION DETAILS

- 4.43.

RELATIVE DEGREE OF DAMAGE OF INADEQUATELY PROTECTED EQUIPMENT

- Minor to major.

FIGURE 3.56. Adequate tank anchorage and flexible line connections such as those shown here increase the operability potential for this fire suppression system.

MOST LIKELY TYPE OR CONSEQUENCE OF DAMAGE FOR
INADEQUATELY PROTECTED EQUIPMENT

- Toppled chemical bottles if bottles are inadequately anchored.
- Ruptured lines if flexible connections are not used.
- Potentially inoperable fire protection system.

Fire Protection Systems

Deluge Equipment

Deluge equipment (Figure 3.57) is required to provide large quantities of water in short order for special hazardous areas where flash fires are likely. All piping and valving should be designed with low flange loads to prevent pipe ruptures. Piping should pass freely from space to space and not be tightly restrained by walls or floors.

EQUIPMENT SEISMIC CATEGORY

- "A" critical equipment.

SEISMIC SPECIFICATION

- SDS-1.

FIGURE 3.57. Example deluge valve. Photograph courtesy of Grinnell Fire Protection Systems Company, Inc.

SEISMIC QUALIFICATION APPROACH

- Equivalent static coefficient analysis.
  - Pipe hangers.
  - Lateral bracing.
- Stress analysis.
  - Flange loads.
- Design team judgment.
  - Provide flexible connections at fixed end joints where possible.

REFERENCE FIGURE FOR INSTALLATION DETAILS

- 4.87, 4.88, 4.89, 4.90, 4.91, 4.92, 4.93, 4.94, 4.95.

RELATIVE DEGREE OF DAMAGE OF INADEQUATELY PROTECTED EQUIPMENT

- Major.

MOST LIKELY TYPE OR CONSEQUENCE OF DAMAGE FOR
INADEQUATELY PROTECTED EQUIPMENT

- Ruptured lines.
- Serious flooding.

- Inoperable fire protection system.
- General cleanup required.

Fire Protection Systems

Detectors, Smoke

Smoke detectors (Figure 3.58) are generally not susceptible to dynamic motions as long as they remain anchored and as long as they retain power.

EQUIPMENT SEISMIC CATEGORY

- "A" critical equipment.

SEISMIC SPECIFICATION

- SDS-1.

SEISMIC QUALIFICATION APPROACH

- Design team judgment.
  - Anchor components.

REFERENCE FIGURE FOR INSTALLATION DETAILS

- 4.46

RELATIVE DEGREE OF DAMAGE OF INADEQUATELY PROTECTED EQUIPMENT

- None to minor.

**MOST LIKELY TYPE OR CONSEQUENCE OF DAMAGE FOR
INADEQUATELY PROTECTED EQUIPMENT**

- Even when inadequately supported, smoke detectors generally remain operational as long as power remains uninterrupted.

FIGURE 3.58. Ceiling installation of smoke detector.

Fire Protection Systems

Extinguishers, Hand-Held

These units (Figure 3.59) come in water and pressurized dry chemical models. They may be mounted directly to the wall or contained in a wall cabinet.

EQUIPMENT SEISMIC CATEGORY

- "A" critical equipment.

SEISMIC SPECIFICATION

- SDS-1.

SEISMIC QUALIFICATION APPROACH

- Design team judgment.
  - Provide positive latch doors and quick release holders.

REFERENCE FIGURES FOR INSTALLATION DETAILS

- 4.44, 4.45.

RELATIVE DEGREE OF DAMAGE OF INADEQUATELY PROTECTED EQUIPMENT

- Minor.

MOST LIKELY TYPE OR CONSEQUENCE OF DAMAGE FOR
INADEQUATELY PROTECTED EQUIPMENT

- Extinguishers fall from wall hooks and cabinets.
- Extinguishers may discharge their contents when they fall.

FIGURE 3.59. Hand-held extinguishers, whether mounted in a cabinet or wall hung, require brackets to keep them in place. This extinguisher is not anchored. The cabinet door does, however, have a positive latch, but it has a glass pane that may well break.

- Cabinet-held extinguishers may fall through glass doors.
- General cleanup required.

REFERENCE FIGURE FOR EXAMPLE OF DAMAGED EQUIPMENT

- 3.163.

Fire Protection Systems

Extinguishers, Hoses

Generally the only effect dynamic motions have on hoses is to cause them to fall out of their cabinets. This is a special problem for cabinets with glass panes. Provision for positive latching doors (Figure 3.60) with plastic panes can keep the hoses securely in place.

EQUIPMENT SEISMIC CATEGORY

- "A" critical equipment.

SEISMIC SPECIFICATION

- SDS-1.

FIGURE 3.60. Hose cabinet with hand-held extinguisher. This cabinet has a positive latch, plastic panes, and a tight friction clip to retain the hose. Photograph courtesy of Ruhnau · Evans Ruhnau · Associates.

SEISMIC QUALIFICATION APPROACH

- Design team judgment.
  - Provide positive latching door.
  - Provide positive catch hose head holder.

RELATIVE DEGREE OF DAMAGE OF INADEQUATELY PROTECTED EQUIPMENT

- Minor.

MOST LIKELY TYPE OR CONSEQUENCE OF DAMAGE FOR
INADEQUATELY PROTECTED EQUIPMENT

- Hose ends up on floor if positive latch doors are not used.
- General cleanup required.

REFERENCE FIGURE FOR EXAMPLE OF DAMAGED EQUIPMENT

- 3.163.

Fire Protection System

Sprinklers

The sprinkler piping must remain attached to the structural system of the building and the sprinkler heads (Figure 3.61) themselves should not be damaged by architectural elements such as swaying suspended ceilings.

EQUIPMENT SEISMIC CATEGORY

- "A" critical equipment.

SEISMIC SPECIFICATION

- SDS-1.

SEISMIC QUALIFICATION APPROACH

- Equivalent static coefficient analysis.
  - Pipe restraints.
- Design team judgment.
  - Protect sprinkler heads from damage caused by adjacent equipment such as suspended ceilings, air ducts, and lights.

REFERENCE FIGURES FOR INSTALLATION DETAILS

- 4.47, 4.87, 4.88, 4.89, 4.90, 4.91, 4.92, 4.93, 4.94, 4.95.

RELATIVE DEGREE OF DAMAGE OF INADEQUATELY PROTECTED EQUIPMENT

- Minor to major.

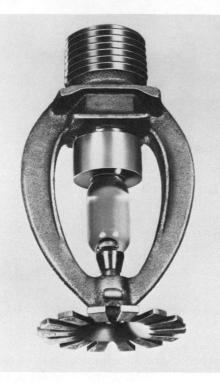

FIGURE 3.61.   Pendant sprinkler. Photograph courtesy of Grinnell Fire Protection Systems Company, Inc.

MOST LIKELY TYPE OR CONSEQUENCE OF DAMAGE FOR
INADEQUATELY PROTECTED EQUIPMENT

- Dislodged water lines.
- Ruptured water lines.
- Facility flooding.
- Inoperable fire system.
- General cleanup required.

Kitchen Systems

Kitchens, especially those found in essential facilities or facilities that could be called on to perform essential functions after an earthquake (schools, etc.), need to continue production in the earthquake aftermath. Failure of some equipment, such as deep fryers, poses a special hazard to kitchen personnel. The Sheet Metal Manufacturers Association of America has produced design guide manuals for equipment that has been reproduced in part in Appendix 3.

SYSTEM SEISMIC CATEGORY

- "C" support equipment.

SYSTEM FOUND IN

- Business establishments.
- Government administration buildings.
- Hospitals.
- Schools.

Kitchen Systems

Deep Fryer

A toppled deep frying unit (Figure 3.62) poses an obvious threat to kitchen personnel.

EQUIPMENT SEISMIC CATEGORY

- "C" support equipment.

FIGURE 3.62. Toppling of deep fryers and the sloshing of hot oil pose a threat unless this equipment is adequately protected. Photograph courtesy of Vulcan-Hart Corporation.

SEISMIC SPECIFICATION

- SDS-2.

SEISMIC QUALIFICATION APPROACH

- Equivalent static coefficient analysis.
  - Base anchorage.
  - Top bracing.
- Design team judgment.
  - SMACNA recommendations.
  - Keep grease unit covered when not in use.

REFERENCE FOR INSTALLATION DETAILS

- Appendix 3.

RELATIVE DEGREE OF DAMAGE OF INADEQUATELY PROTECTED EQUIPMENT

- Minor to moderate.

MOST LIKELY TYPE OR CONSEQUENCE OF DAMAGE FOR
INADEQUATELY PROTECTED EQUIPMENT

- Displaced equipment.
- Toppled equipment.
- Grease spills from sloshing or toppling.
- Potential for severe burns.

Kitchen Systems

Dishwashers

Toppled dishwashers can cause flooding in the kitchen from broken water lines. Figure 3.63 illustrates a typical cabinet type dishwasher.

EQUIPMENT SEISMIC CATEGORY

- "C" support equipment.

SEISMIC SPECIFICATION

- SDS-2.

SEISMIC QUALIFICATION APPROACH

- Equivalent static coefficient analysis.
  - Base anchorage.
- Design team judgment.
  - Flexible water line connections.

FIGURE 3.63. Example dishwashing unit. Photograph courtesy of Vulcan-Hart Corporation.

REFERENCE FOR INSTALLATION DETAILS

● Appendix 3.

RELATIVE DEGREE OF DAMAGE OF INADEQUATELY PROTECTED EQUIPMENT

● Minor to major.

MOST LIKELY TYPE OR CONSEQUENCE OF DAMAGE FOR INADEQUATELY PROTECTED EQUIPMENT

● Potential for flooding from broken water lines.
● Toppled equipment.
● General cleanup required.

Kitchen Systems

Food Mixers

Base anchorage of food mixers (Figure 3.64) reduces the potential for toppled equipment spills. Dislodged mixing bowls can also cause spills.

EQUIPMENT SEISMIC CATEGORY

● "C" support equipment.

FIGURE 3.64. Food mixer with base anchorage provisions. Photograph courtesy of Vulcan-Hart Corporation.

SEISMIC SPECIFICATION

- SDS-2.

SEISMIC QUALIFICATION APPROACH

- Equivalent static coefficient analysis.
  - Base anchorage.

REFERENCE FOR INSTALLATION DETAILS

- Appendix 3.

RELATIVE DEGREE OF DAMAGE OF INADEQUATELY PROTECTED EQUIPMENT

- Minor to moderate.

MOST LIKELY TYPE OR CONSEQUENCE OF DAMAGE FOR
INADEQUATELY PROTECTED EQUIPMENT

- Shifted mixers.
- Toppled mixers.
- Spilled food.
- General cleanup required.

Kitchen Systems

Kettles, Console Construction

Console-type kettles (Figure 3.65), even though large, can be dislodged and should be anchored. Kettle covers should be employed to reduce spillage.

EQUIPMENT SEISMIC CATEGORY

- "C" support equipment.

SEISMIC SPECIFICATION

- SDS-2.

SEISMIC QUALIFICATION APPROACH

- Design team judgment.
  - Provide nonpermanent base anchorage if the unit is to remain portable.
  - Flexible service line connections.
- Equivalent static coefficient analysis.
  - Provide permanent base anchorage if the equipment does not need to be portable.

REFERENCE FOR INSTALLATION DETAILS

- Appendix 3.

RELATIVE DEGREE OF DAMAGE OF INADEQUATELY PROTECTED EQUIPMENT

- Minor to moderate.

MOST LIKELY TYPE OR CONSEQUENCE OF DAMAGE FOR
INADEQUATELY PROTECTED EQUIPMENT

- Dislodged equipment.
- Spilled contents.
- General cleanup required.

FIGURE 3.65.   Wall-mounted kettles. Photograph courtesy of Groen, a division of the Dover Corporation.

Kitchen Systems

Kettles, Steam-Jacketed

If kettles (see Figure 3.66) are not anchored, they are likely to topple, spilling their contents and possibly rupturing their gas supply lines.

EQUIPMENT SEISMIC CATEGORY

● "C" support equipment.

SEISMIC SPECIFICATION

● SDS-2.

SEISMIC QUALIFICATION APPROACH

● Equivalent static coefficient analysis.
   • Base anchorage for units that are provided with drilled base plates by the manufacturer.

FIGURE 3.66. Gas-heated, self-contained, steam-jacketed kettle. Photograph courtesy of Groen, A Division of the Dover Corporation.

- Design team judgment.
  - Provide flexible gas and electric connections.
  - Provide base anchorage where base plates are not provided by the manufacturer.

REFERENCE FOR INSTALLATION DETAILS

- Appendix 3.

RELATIVE DEGREE OF DAMAGE OF INADEQUATELY PROTECTED EQUIPMENT

- Minor to moderate.

MOST LIKELY TYPE OR CONSEQUENCE OF DAMAGE FOR
INADEQUATELY PROTECTED EQUIPMENT

- Toppled kettles.
- Spilled food.
- General cleanup required.

Kitchen Systems

Portable Tray Carts, Temperature Controlled

Temperature controlled food carts (Figure 3.67) are relatively expensive and if not protected are highly susceptible to damage. Wheel locks do not prevent toppling. These units should be restrained when not in use.

EQUIPMENT SEISMIC CATEGORY

- ''D'' support equipment.

SEISMIC SPECIFICATION

- SDS-2.

SEISMIC QUALIFICATION APPROACH

- Design team judgment.
  - Provide anchorage for equipment when it is not in use.

REFERENCE FOR INSTALLATION DETAILS

- Appendix 3.

RELATIVE DEGREE OF DAMAGE OF INADEQUATELY PROTECTED EQUIPMENT

- Minor.

MOST LIKELY TYPE OR CONSEQUENCE OF DAMAGE FOR
INADEQUATELY PROTECTED EQUIPMENT

- Dislodged equipment.
- Toppled equipment.

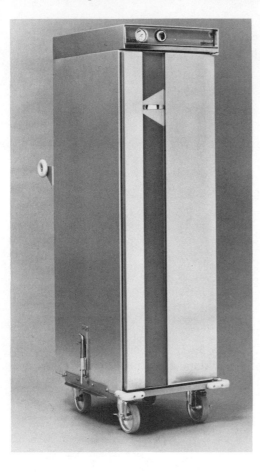

FIGURE 3.67. Temperature controlled portable tray cart. Note the anchorage lever that could be used for restrained storage of the cart when it is not in use. Photograph courtesy of Precision Metal Products, Incorporated.

- Spilled food, and so on.
- General cleanup required.

Kitchen Systems

Refrigerators, Large Units

Very large objects such as refrigerators (see Figure 3.68) can also be dislodged by an earthquake if not adequately protected.

EQUIPMENT SEISMIC CATEGORY

- "C" support equipment.

SEISMIC SPECIFICATION

- SDS-2.

FIGURE 3.68.   Large walk-in refrigerator unit. Photograph courtesy of Vulcan-Hart Corporation.

SEISMIC QUALIFICATION APPROACH

- Equivalent static coefficient analysis.
  - Anchorage of units.
- Design team judgment.
  - Provide shelf-lip restraints to prevent unnecessary spills.

REFERENCE FOR INSTALLATION DETAILS

- Appendix 3.

RELATIVE DEGREE OF DAMAGE OF INADEQUATELY PROTECTED EQUIPMENT

- Minor to moderate.

MOST LIKELY TYPE OR CONSEQUENCE OF DAMAGE FOR
INADEQUATELY PROTECTED EQUIPMENT

- Dislodged equipment.
- Spilled items.
- General cleanup required.

Kitchen Systems

Refrigerators, Medium (Figure 3.69) and Small Units

All tall and slender equipment should be considered for the earthquake environment.

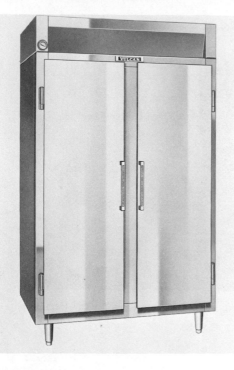

FIGURE 3.69. Medium sized kitchen refrigerator. Photograph courtesy of Vulcan-Hart Corporation.

EQUIPMENT SEISMIC CATEGORY

- "C" support equipment.

SEISMIC SPECIFICATION

- SDS-2.

SEISMIC QUALIFICATION APPROACH

- Equivalent static coefficient analysis.
  - Base anchorage.
  - Top bracing.

REFERENCE FOR INSTALLATION DETAILS

- Appendix 3.

RELATIVE DEGREE OF DAMAGE OF INADEQUATELY PROTECTED EQUIPMENT

- Minor to moderate.

MOST LIKELY TYPE OR CONSEQUENCE OF DAMAGE FOR INADEQUATELY PROTECTED EQUIPMENT

- Shifted refrigerators.
- Toppled refrigerators.

- Inoperative refrigerators.
- Potential for personnel injury.
- General cleanup required.

Kitchen Systems

Serving Lines

High speed makeup systems (Figure 3.70) can be wheeled or stationary. The individual pieces of equipment should be tied together to reduce the toppling potential. Stationary equipment should be base anchored.

EQUIPMENT SEISMIC CATEGORY

- "D" or "E" miscellaneous equipment.

SEISMIC SPECIFICATION

- SDS-2.

SEISMIC QUALIFICATION APPROACH

- Equivalent static coefficient analysis.
  - Base anchorage for stationary equipment.
- Design team judgment.
  - Connect wheeled modules together to reduce toppling potential.
  - Consider modular anchorage system for wheeled equipment.

REFERENCE FOR INSTALLATION DETAILS

- Appendix 3.

RELATIVE DEGREE OF DAMAGE OF INADEQUATELY PROTECTED EQUIPMENT

- Minor to moderate.

FIGURE 3.70. High-speed makeup serving line. Photograph courtesy of Precision Metal Products, Incorporated.

MOST LIKELY TYPE OR CONSEQUENCE OF DAMAGE FOR
INADEQUATELY PROTECTED EQUIPMENT

- Dislodged equipment.
- Toppled equipment.
- Spilled food, boiling water, dishes, and so on.
- General cleanup required.

Kitchen Systems

Steam Cookers

Anchorage of this type of equipment (Figure 3.71) increases the operability potential of kitchens found in essential facilities such as hospitals.

EQUIPMENT SEISMIC CATEGORY

- "C" support equipment.

SEISMIC SPECIFICATION

- SDS-2.

SEISMIC QUALIFICATION APPROACH

- Equivalent static coefficient analysis.
  - Base anchorage.

FIGURE 3.71.   Upright steam cooker. Photograph courtesy of Vulcan-Hart Corporation.

- Design team judgment.
  - Flexible electrical connections.
  - Flexible gas connections.

REFERENCE FOR INSTALLATION DETAILS

- Appendix 3.

RELATIVE DEGREE OF DAMAGE OF INADEQUATELY PROTECTED EQUIPMENT

- Minor to moderate.

MOST LIKELY TYPE OR CONSEQUENCE OF DAMAGE FOR
INADEQUATELY PROTECTED EQUIPMENT

- Dislodged equipment.
- Broken electrical connections.
- Broken gas lines.
- Toppled equipment.

Kitchen Systems

Stoves and Ovens, Large (Modular Applications)

Individual pieces of equipment can be hooked together into modular applications (Figure 3.72) that may not be as likely to topple as the individual units, but nonetheless require earthquake protection.

EQUIPMENT SEISMIC CATEGORY

- "C" support equipment.

SEISMIC SPECIFICATION

- SDS-2.

SEISMIC QUALIFICATION APPROACH

- Equivalent static coefficient analysis.
  - Base anchorage.
- Design team judgment.
  - Flexible gas and electrical connections between the building structure and the equipment.

REFERENCE FOR INSTALLATION DETAILS

- Appendix 3.

RELATIVE DEGREE OF DAMAGE OF INADEQUATELY PROTECTED EQUIPMENT

- Minor to moderate.

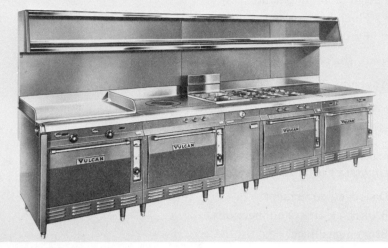

FIGURE 3.72. Modular cook-top assembly. Photograph courtesy of Vulcan-Hart Corporation.

MOST LIKELY TYPE OR CONSEQUENCE OF DAMAGE FOR
INADEQUATELY PROTECTED EQUIPMENT

- Dislodged equipment.
- Spilled grease, and so on.
- Potential for personnel injury.
- General cleanup required.

Kitchen Systems

Stoves and Ovens, Small (Figure 3.73)

Toppled or dislodged stoves can injure personnel as well as create significant cleanup problems.

EQUIPMENT SEISMIC CATEGORY

- "C" support equipment.

SEISMIC SPECIFICATION

- SDS-2.

SEISMIC QUALIFICATION APPROACH

- Equivalent static coefficient analysis.
  - Base anchorage.
- Design team judgment.
  - Flexible gas and electrical connections between the building structure and the equipment.

FIGURE 3.73. Small range. Photograph courtesy of Vulcan-Hart Corporation.

REFERENCE FOR INSTALLATION DETAILS

- Appendix 3.

RELATIVE DEGREE OF DAMAGE OF INADEQUATELY PROTECTED EQUIPMENT

- Minor to moderate.

MOST LIKELY TYPE OR CONSEQUENCE OF DAMAGE FOR
INADEQUATELY PROTECTED EQUIPMENT

- Dislodged stoves and ovens.
- Toppled stoves and ovens.
- Spilled grease, and so on.
- Ruptured gas lines and electrical connections.
- General cleanup required.

Kitchen Systems

Urns/Brewers

Urns that are toppled as a result of an earthquake can spill scalding water.

EQUIPMENT SEISMIC CATEGORY

- "E" miscellaneous equipment.

SEISMIC SPECIFICATION

- SDS-2.

SEISMIC QUALIFICATION APPROACH

- Equivalent static coefficient analysis.
  - Base anchorage for urns (Figure 3.74) that are provided with drilled base plates by manufacturers.
- Design team judgment.
  - Provide flexible electrical and gas lines.
  - Provide base anchorage for urns that are not provided with base plates by manufacturers.

REFERENCE FOR INSTALLATION DETAILS

- Appendix 3.

RELATIVE DEGREE OF DAMAGE OF INADEQUATELY PROTECTED EQUIPMENT

- Minor.

MOST LIKELY TYPE OR CONSEQUENCE OF DAMAGE FOR INADEQUATELY PROTECTED EQUIPMENT

- Toppled urns.
- Spilled contents.
- Potential for personnel injury.
- General cleanup required.

FIGURE 3.74. Urn with base anchorage provisions. Photograph courtesy of Groen, A Division of the Dover Corporation.

Kitchen Systems

Utility Carts

Whether wheeled or stationary, utility carts (Figure 3.75) are highly susceptible to earthquakes. Even if the carts do not topple, items stored on them
are commonly spilled. Wheel locks do not prevent toppling or spilled equipment.

EQUIPMENT SEISMIC CATEGORY

● "D" or "E" miscellaneous equipment.

SEISMIC SPECIFICATION

● SDS-2.

SEISMIC QUALIFICATION APPROACH

● Equivalent static coefficient analysis.
  • Base anchorage and top bracing for stationary equipment.
● Design team judgment.
  • Restraints for wheeled equipment when not in use.
  • Where practical, parapets or shelf restraints for shelved items.

REFERENCE FOR INSTALLATION DETAILS

● Appendix 3.

FIGURE 3.75. Examples of various
utility carts and racks. Photograph
courtesy of Precision Metal Products,
Incorporated.

RELATIVE DEGREE OF DAMAGE OF INADEQUATELY PROTECTED EQUIPMENT

● Minor.

MOST LIKELY TYPE OR CONSEQUENCE OF DAMAGE FOR
INADEQUATELY PROTECTED EQUIPMENT

● Toppled carts.
● Spilled dishes, food, and so on.
● General cleanup required.

Kitchen Systems

Ventilators

Overhead ventilators (Figure 3.76) are generally attached to the building structure. Their anchorage should conform to the expected earthquake environment to prevent their collapse.

EQUIPMENT SEISMIC CATEGORY

● "C" support equipment.

SEISMIC SPECIFICATION

● SDS-2.

SEISMIC QUALIFICATION APPROACH

● Equivalent static coefficient analysis.
  • Anchorage to the building structure.
● Design team judgment.
  • Review fasteners for louvers, and so on.

REFERENCE FOR INSTALLATION DETAILS

● Appendix 3.

FIGURE 3.76.   Ventilating unit. Photograph courtesy of Gaylord Industries, Inc.

RELATIVE DEGREE OF DAMAGE OF INADEQUATELY PROTECTED EQUIPMENT

- Minor to major.

MOST LIKELY TYPE OR CONSEQUENCE OF DAMAGE FOR
INADEQUATELY PROTECTED EQUIPMENT

- Dislodged hoods.
- Dislodged vent plates.
- Potential for personnel injury.
- General cleanup required.

Kitchen Systems

Wheeled Equipment

Wheel locks (Figure 3.77), if set, keep equipment from extensive rolling. They do not, however, prevent sliding or toppling.

EQUIPMENT SEISMIC CATEGORY

- ''D'' support equipment.

SEISMIC SPECIFICATION

- SDS-2.

SEISMIC QUALIFICATION APPROACH

- Design team judgment.
  - Provide modular anchors for all wheeled equipment.

REFERENCE FOR INSTALLATION DETAILS

- Appendix 3.

FIGURE 3.77. Portable steam table. Photograph courtesy of Precision Metal Products, Incorporated.

RELATIVE DEGREE OF DAMAGE OF INADEQUATELY PROTECTED EQUIPMENT

- Minor to moderate.

MOST LIKELY TYPE OR CONSEQUENCE OF DAMAGE FOR
INADEQUATELY PROTECTED EQUIPMENT

- Rolling equipment—potential for collision.
- Toppled equipment.
- Spilled grease, and so on.
- Potential for personnel injury.
- General cleanup required.

Lighting Systems

Most facilities can only operate on a limited basis if general lighting failures should occur. Because of its geometry, fluorescent lighting generally tends to be more vulnerable to earthquakes than incandescent lighting.

SYSTEM SEISMIC CATEGORY

- ''B'' support equipment.

SYSTEM FOUND IN

- All building types.

Lighting Systems

Emergency Lights

All essential facilities require emergency light capabilities. Other types of required light capabilities exist, depending on the type of facility. Emergency lights (see Figure 3.78) are all too often left unanchored on shelves, and so on.

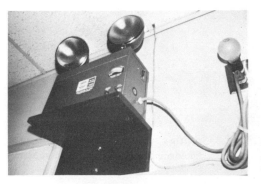

FIGURE 3.78. Adequately protected emergency light. The light is anchored to the wall bracket.

EQUIPMENT SEISMIC CATEGORY

- "A" critical equipment.

SEISMIC SPECIFICATION

- SDS-1.

SEISMIC QUALIFICATION APPROACH

- Equivalent static coefficient analysis.
  - Anchorage of bracket to wall.
  - Anchorage of emergency light to bracket.

REFERENCE FIGURE FOR INSTALLATION DETAILS

- 4.48.

RELATIVE DEGREE OF DAMAGE OF INADEQUATELY PROTECTED EQUIPMENT

- Moderate.

MOST LIKELY TYPE OR CONSEQUENCE OF DAMAGE FOR INADEQUATELY PROTECTED EQUIPMENT

- If not correctly supported, emergency lights will fall from their perches.
- Lights are likely to be inoperative if they fall.
- General cleanup required.

Lighting Systems

Fluorescent Lighting

Fluorescent lighting (Figure 3.79) is especially susceptible to dynamic motions. Suspended fixtures sway wildly, diffusers fall, and tubes dislodge from their holders.

FIGURE 3.79. Fluorescent lighting mounted in a T-bar ceiling. Photograph courtesy of Ruhnau · Evans · Ruhnau · Associates.

EQUIPMENT SEISMIC CATEGORY

- "B" support equipment.

SEISMIC SPECIFICATION

- SDS-1.

SEISMIC QUALIFICATION APPROACH

- Design team judgment.
  - Provide for bracing and safety wires.
- Equivalent static coefficient analysis.
  - For main supports and bracing.
- Dynamic analysis.
  - If swaying is likely to preclude possible collisions.

REFERENCE FIGURES FOR INSTALLATION DETAILS

- 4.49, 4.50, 4.51.

RELATIVE DEGREE OF DAMAGE OF INADEQUATELY PROTECTED EQUIPMENT

- Moderate to major.

MOST LIKELY TYPE OR CONSEQUENCE OF DAMAGE FOR
INADEQUATELY PROTECTED EQUIPMENT

- Collapse of lighting fixtures.
- Collapse of fixture subcomponents (tubes, diffusers, etc.).
- Collision of swaying fixtures.
- Potentially inoperative lights.
- Potential for personnel injury.
- General cleanup required.

REFERENCE FIGURES FOR EXAMPLE OF DAMAGED EQUIPMENT

- 3.165, 3.166, 3.167.

Lighting Systems

Incandescent Lighting

If fixed, incandescent lighting (Figure 3.80) generally is not susceptible to damage. Suspended lights, unless restrained, may sway violently with the potential for collision with other lights, walls, and so on.

EQUIPMENT SEISMIC CATEGORY

- "B" support equipment.

SEISMIC SPECIFICATION

- SDS-1.

FIGURE 3.80. Incandescent ceiling lighting. Photograph courtesy of Ruhnau · Evans · Ruhnau · Associates.

SEISMIC QUALIFICATION APPROACH

● Design team judgment.
 • For lightweight units.
● Equivalent static coefficient analysis.
 • For heavier units.
● Dynamic analysis.
 • To determine possible collisions if swaying cannot be avoided.

REFERENCE FIGURES FOR INSTALLATION DETAILS

● 4.49, 4.50, 4.51, 4.52.

RELATIVE DEGREE OF DAMAGE OF INADEQUATELY PROTECTED EQUIPMENT

● Minor.

MOST LIKELY TYPE OR CONSEQUENCE OF DAMAGE FOR
INADEQUATELY PROTECTED EQUIPMENT

● Dislodged fixtures.
● Possibly inoperative.

REFERENCE FIGURE FOR EXAMPLE OF DAMAGED EQUIPMENT

● 3.168.

Lighting Systems

Table and Desk Lighting

Table lamps such as reading lamps can be expected to fall unless restrained. Drafting lamps (Figure 3.81) will likely sway.

EQUIPMENT SEISMIC CATEGORY

● "E" miscellaneous equipment.

FIGURE 3.81. Drafting lamps such as those shown here are likely to sway violently unless restrained. Photograph courtesy of Ruhnau • Evans • Ruhnau • Associates.

SEISMIC SPECIFICATION

- SDS-2.

SEISMIC QUALIFICATION APPROACH

- Design team judgment.
  - Provide base anchorage if possible.
  - Provide swing arm restraint.

REFERENCE FIGURE FOR INSTALLATION DETAILS

- 4.52.

RELATIVE DEGREE OF DAMAGE OF INADEQUATELY PROTECTED EQUIPMENT

- Minor.

MOST LIKELY TYPE OR CONSEQUENCE OF DAMAGE FOR
INADEQUATELY PROTECTED EQUIPMENT

- Upset lamps.
- General cleanup required.

Medical Systems

Medical systems vary from complex critical and life support equipment to ad hoc storage. Seismic qualification programs for medical facilities must never be neglected in earthquake prone areas. Hospitals and medical clinics are facilities likely to suffer the greatest adverse effects of an earthquake. They have daily duties that keep personnel from maintaining an ever-present guard against potential aseismic deficiencies. Therefore, it is the design team's charge to provide a mechanism whereby hospital equipment is automatically protected from damaging earthquakes whenever possible. If this is not done, after an earthquake we can expect an essential facility that is unable to function because much of its equipment is left lying broken and twisted on the floor after the shaking stops, which is certainly not conducive to performing essential functions.

SYSTEM SEISMIC CATEGORY

● Varies with subsystems—critical to miscellaneous systems.

SYSTEM FOUND IN

● Hospitals/clinics.

Medical Systems

Anesthesia Cart

Anesthesia carts (Figure 3.82) are highly complex pieces of equipment that are sensitive to earthquake motions and can cause adverse secondary effects should they be damaged.

EQUIPMENT SEISMIC CATEGORY

● "A" critical equipment.

SEISMIC SPECIFICATION

● SDS-1.

FIGURE 3.82. Unrestrained operating room anesthesia cart.

SEISMIC QUALIFICATION APPROACH

- Design team judgment.
  - Protect cart from dynamic motions.
- Equivalent static coefficient analysis.
  - For restraining system when equipment is not in use.

REFERENCE FIGURE FOR INSTALLATION DETAILS

- 4.69.

RELATIVE DEGREE OF DAMAGE OF INADEQUATELY PROTECTED EQUIPMENT

- Moderate to major.

MOST LIKELY TYPE OR CONSEQUENCE OF DAMAGE FOR
INADEQUATELY PROTECTED EQUIPMENT

- Toppled equipment.
- Runaway equipment.
- Spilled chemicals.
- Personnel injury possible.
- Potentially inoperable equipment.
- General cleanup required.

Medical Systems

Bacteriology Test Equipment

Damaged bacteriology equipment could possibly contaminate entire sections of the hospital at a time when complications are definitely not needed. Very careful consideration should be given to all such equipment (see Figure 3.83).

EQUIPMENT SEISMIC CATEGORY

- "C" support equipment.

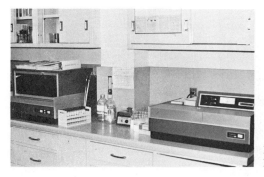

FIGURE 3.83. Typical   bacteriology laboratory scene.

SEISMIC SPECIFICATION

- SDS-2.

SEISMIC QUALIFICATION APPROACH

- Equivalent static coefficient analysis.
  - Base anchorage.
- Design team judgment.
  - Consider all bacteriology laboratory equipment to prevent the potential of contamination.

REFERENCE FIGURES FOR INSTALLATION DETAILS

- 4.53, 4.54, 4.55, 4.56, 4.57, 4.59, 4.60, 4.61, 4.62, 4.63, 4.70, 4.71, 4.72, 4.73, 4.76, 4.77.

RELATIVE DEGREE OF DAMAGE OF INADEQUATELY PROTECTED EQUIPMENT

- Minor to major (should contamination occur).

MOST LIKELY TYPE OR CONSEQUENCE OF DAMAGE FOR
INADEQUATELY PROTECTED EQUIPMENT

- Toppled equipment.
- Hospital contamination possible.
- Inoperative test equipment.
- General cleanup and/or decontamination required.

REFERENCE FIGURES FOR EXAMPLES OF DAMAGED EQUIPMENT

- 3.170, 3.171.

Medical Systems

Blood Bank Refrigerator

Blood bank refrigerators (Figure 3.84) are tall, slender pieces of equipment with substantial weight and are commonly located within a work space. Toppling could cause personnel injury.

EQUIPMENT SEISMIC CATEGORY

- "A" critical equipment.

SEISMIC SPECIFICATION

- SDS-1.

SEISMIC QUALIFICATION APPROACH

- Equivalent static coefficient analysis.

FIGURE 3.84. Unsecured blood bank refrigerator in the laboratory work space.

- • Base anchorage.
- • Top anchorage (do not anchor to weak walls, etc.).
- ● Design team judgment.
  - • Provide positive latch doors (magnetic and friction catches do not work well in an earthquake) plus drawer stops.

REFERENCE FIGURES FOR INSTALLATION DETAILS

- ● 4.70, 4.71.

RELATIVE DEGREE OF DAMAGE OF INADEQUATELY PROTECTED EQUIPMENT

- ● Moderate.

MOST LIKELY TYPE OR CONSEQUENCE OF DAMAGE FOR INADEQUATELY PROTECTED EQUIPMENT

- ● Toppled equipment.
- ● Spilled items if positive latches are not used.
- ● Possible personnel injury.
- ● General cleanup required.

Medical Systems

Blood Chemistry Analyzer

The console-type blood chemistry analyzer (Figure 3.85) is quite an expensive piece of equipment that should not be damaged needlessly. Adequate protection can be provided through a seismic qualification program.

EQUIPMENT SEISMIC CATEGORY

- "C" support equipment.

SEISMIC SPECIFICATION

- SDS-2.

SEISMIC QUALIFICATION APPROACH

- Equivalent static coefficient analysis.
  - Base anchorage.
- Manufacturers may wish to
  - Dynamically analyze console frame.
  - Perform generic test for equipment operability.

REFERENCE FIGURE FOR INSTALLATION DETAILS

- 4.58.

RELATIVE DEGREE OF DAMAGE OF INADEQUATELY PROTECTED EQUIPMENT

- Minor to moderate.

MOST LIKELY TYPE OR CONSEQUENCE OF DAMAGE FOR
INADEQUATELY PROTECTED EQUIPMENT

- Dislodged equipment.
- Toppled equipment.

FIGURE 3.85. This blood chemistry analyzer is likely to be damaged during an earthquake because it has not been anchored.

- Potentially inoperative analyzer.
- General cleanup required.

Medical Systems

Blood Chemistry Analyzer, Counter Top Model

The counter top blood chemistry analyzer shown in Figure 3.86 can easily topple and possibly injure the operator seated adjacent to the equipment.

EQUIPMENT SEISMIC CATEGORY

- "C" support equipment.

SEISMIC SPECIFICATION

- SDS-2.

SEISMIC QUALIFICATION APPROACH

- Equivalent static coefficient analysis.
  - Base anchorage must be provided to protect equipment such as this, as well as personnel.

REFERENCE FIGURES FOR INSTALLATION DETAILS

- 4.60, 4.61, 4.62, 4.63.

RELATIVE DEGREE OF DAMAGE OF INADEQUATELY PROTECTED EQUIPMENT

- Moderate to major.

MOST LIKELY TYPE OR CONSEQUENCE OF DAMAGE FOR
INADEQUATELY PROTECTED EQUIPMENT

- Toppled equipment.
- Potential for personnel injury.
- Inoperative equipment.
- General cleanup required.

FIGURE 3.86. Personnel working adjacent to this small tabletop analyzer could be injured if it should fall.

Medical Systems

Blood Chemistry Coagulyzer

Tabletop coagulyzers used in blood chemistry laboratories require base anchorage so that the equipment does not slide off the counter top onto the floor.

EQUIPMENT SEISMIC CATEGORY

- "C" support equipment.

SEISMIC SPECIFICATION

- SDS-2.

SEISMIC QUALIFICATION APPROACH

- Equivalent static coefficient analysis.
  - Base anchorage.
- Design team judgment.
  - For other restraining methods.

REFERENCE FIGURES FOR INSTALLATION DETAILS

- 4.60, 4.61, 4.62, 4.63.

RELATIVE DEGREE OF DAMAGE OF INADEQUATELY PROTECTED EQUIPMENT

- Moderate to major.

MOST LIKELY TYPE OR CONSEQUENCE OF DAMAGE FOR
INADEQUATELY PROTECTED EQUIPMENT

- Toppled coagulyzer.
- Possibly inoperative coagulyzer.
- General cleanup required.

FIGURE 3.87. Tabletop coagulizers such as that shown here require consideration.

Medical Systems

Centrifuge, Large Portable

Although not a critical piece of equipment, heavy wheeled equipment such as the centrifuge (Figure 3.88) can be the cause of significant damage once set into motion by an earthquake.

EQUIPMENT SEISMIC CATEGORY

● "D" support equipment.

SEISMIC SPECIFICATION

● SDS-2.

SEISMIC QUALIFICATION APPROACH

● Equivalent static coefficient analysis.

REFERENCE FIGURE FOR INSTALLATION DETAILS

● 4.70.

FIGURE 3.88. Example of a large portable unsecured centrifuge.

RELATIVE DEGREE OF DAMAGE OF INADEQUATELY PROTECTED EQUIPMENT

● Possibly major (especially to other equipment).

MOST LIKELY TYPE OR CONSEQUENCE OF DAMAGE FOR
INADEQUATELY PROTECTED EQUIPMENT

● Runaway centrifuge.
● Collision with adjacent equipment.
● Toppled equipment.
● Inoperative equipment.
● General cleanup required.

Medical Systems

Centrifuge, Small Portable

The lightweight centrifuges (Figure 3.89) require the same consideration as all other counter top equipment.

EQUIPMENT SEISMIC CATEGORY

● "C" support equipment.

SEISMIC SPECIFICATION

● SDS-2.

SEISMIC QUALIFICATION APPROACH

● Design team judgment.
  • Equipment restraints to keep the equipment on the counter top.

REFERENCE FIGURES FOR INSTALLATION DETAILS

● 4.60, 4.61, 4.62, 4.63.

FIGURE 3.89. Unsecured tabletop centrifuges (white cylindrical objects on the counter top).

RELATIVE DEGREE OF DAMAGE OF INADEQUATELY PROTECTED EQUIPMENT

● Minor.

MOST LIKELY TYPE OR CONSEQUENCE OF DAMAGE OF
INADEQUATELY PROTECTED EQUIPMENT

● Equipment sliding off counter top.
● Potentially inoperative centrifuges.
● General cleanup required.

Medical Systems

Chemical Storage

Using plastic bottles (Figure 3.90) for wet and dry chemical storage greatly decreases cleanup requirements and prevents potential adverse reactions from mixing chemicals if they spill on the floor.

EQUIPMENT SEISMIC CATEGORY

● "C" support equipment.

FIGURE 3.90.   Plastic bottle chemical storage.

SEISMIC SPECIFICATION

- SDS-2.

SEISMIC QUALIFICATION APPROACH

- Design team judgment.
  - Decision to use plastic bottles lies with hospital administration.
  - Provide tilted shelves and shelf restraints to help prevent spills.

REFERENCE FIGURES FOR INSTALLATION DETAILS

- 4.53, 4.54, 4.55, 4.56, 4.57, 4.70, 4.71, 4.73, 4.74.

RELATIVE DEGREE OF DAMAGE OF INADEQUATELY PROTECTED EQUIPMENT

- Minor.

MOST LIKELY TYPE OR CONSEQUENCE OF DAMAGE FOR
INADEQUATELY PROTECTED EQUIPMENT

- Broken glass if plastic bottles not used.
- Adverse chemical reactions are possible.
- General cleanup of broken and unbroken containers if shelf restraint schemes are not employed.

REFERENCE FIGURE FOR EXAMPLE OF DAMAGED EQUIPMENT

- 3.169.

Medical Systems

Chemotherapy Treatment

Chemotherapy treatment systems (Figure 3.91), although not critical to hospital operations, are very expensive, heavy, and worth extensive protection programs.

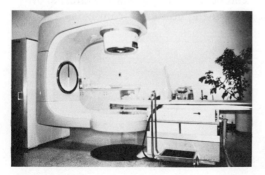

FIGURE 3.91. Large   eccentrically loaded chemotherapy unit.

EQUIPMENT SEISMIC CATEGORY

- "C" support equipment.

SEISMIC SPECIFICATION

- SDS-2.

SEISMIC QUALIFICATION APPROACH

- Equivalent static coefficient analysis.
  - Base anchorage.
- Dynamic analysis or seismic test.
  - Manufacturers may wish to undertake more ambitious generic qualification programs to protect their equipment from operation failures and subsequent expensive repairs.

REFERENCE FIGURE FOR INSTALLATION DETAILS

- 4.75.

RELATIVE DEGREE OF DAMAGE OF INADEQUATELY PROTECTED EQUIPMENT

- Minor—because of its geometry, this equipment is generally anchored.

MOST LIKELY TYPE OR CONSEQUENCE OF DAMAGE FOR
INADEQUATELY PROTECTED EQUIPMENT

- Potential operational failure.
- Base anchorage failure if underdesigned.
- General cleanup required.

Medical Systems

Cold Storage

Cold storage equipment (Figure 3.92), whether it be for the Pathology Laboratory or for pharmaceuticals, is generally tall, slender, and heavy. Doors need positive latches, not magnetic catches. Rolling or sliding internal drawers should have stops and lips to prevent the contents from being spilled.

EQUIPMENT SEISMIC CATEGORY

- "C" support equipment.

SEISMIC SPECIFICATION

- SDS-2.

SEISMIC QUALIFICATION APPROACH

- Equivalent static coefficient analysis.
  - Base and top anchorage.

FIGURE 3.92. Unsecured example of a cold storage unit.

- Design team judgment.
  - Provide positive latch doors and drawer stops.

REFERENCE FIGURES FOR INSTALLATION DETAILS

- 4.70, 4.71.

RELATIVE DEGREE OF DAMAGE OF INADEQUATELY PROTECTED EQUIPMENT

- Minor to moderate.

MOST LIKELY TYPE OR CONSEQUENCE OF DAMAGE FOR
INADEQUATELY PROTECTED EQUIPMENT

- Toppled storage unit.
- Spilled contents.
- General cleanup required.

Medical Systems

Crash Cart

Crash carts (Figure 3.93) contain essential medicines and tools for in-hospital emergency treatment.

FIGURE 3.93. Wheel-mounted crash cart without anchorage provisions. This vital piece of equipment is likely to be upset during an earthquake.

EQUIPMENT SEISMIC CATEGORY

- "A" critical equipment.

SEISMIC SPECIFICATION

- SDS-1.

SEISMIC QUALIFICATION APPROACH

- Design team judgment.
  - Design restraining system for times when equipment is not in use.
- Equivalent static coefficient analysis.
  - Anchorage of restraining system.

REFERENCE FIGURE FOR INSTALLATION DETAILS

- 4.69.

RELATIVE DEGREE OF DAMAGE OF INADEQUATELY PROTECTED EQUIPMENT

- Minor to moderate.

MOST LIKELY TYPE OR CONSEQUENCE OF DAMAGE FOR INADEQUATELY PROTECTED EQUIPMENT

- Toppled carts.
- Runaway carts.

- Collision with other equipment.
- Potential for personnel injury.
- General cleanup required.

Medical Systems

Culture Incubator, Counter Top

Counter top culture incubators (Figure 3.94) require anchorage to prevent their toppling to the floor.

EQUIPMENT SEISMIC CATEGORY

- "C" support equipment.

SEISMIC SPECIFICATION

- SDS-2.

SEISMIC QUALIFICATION APPROACH

- Design team judgment.
  - Provide for adequate installation space.
  - Use positive latch doors; magnetic catches and some friction catches do not work well in earthquakes (the doors open and the contents spill out).
- Equivalent static coefficient analysis.
  - Anchorage of counter.
  - Anchorage of incubator to counter.

FIGURE 3.94.   Unsecured example of a culture incubator atop a small counter.

REFERENCE FIGURES FOR INSTALLATION DETAILS

● 4.60, 4.61, 4.62, 4.63.

RELATIVE DEGREE OF DAMAGE OF INADEQUATELY PROTECTED EQUIPMENT

● Minor to moderate.

MOST LIKELY TYPE OR CONSEQUENCE OF DAMAGE FOR
INADEQUATELY PROTECTED EQUIPMENT

● Toppled equipment.
● Possibly inoperative incubator.
● Spilled contents.
● General cleanup required.

Medical Systems

Exam Equipment, Specialized

This equipment (Figure 3.95) is generally needed for regular patients on a day-to-day basis, but nonetheless deserves consideration for its own protection.

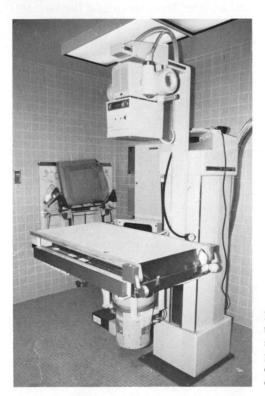

FIGURE 3.95. Specialized examination equipment such as the urology table shown here typically has geometries that produce extreme motions during an earthquake. Note the base anchorage employed on this equipment.

EQUIPMENT SEISMIC CATEGORY

- "C" support equipment.

SEISMIC SPECIFICATION

- SDS-2.

SEISMIC QUALIFICATION APPROACH

- Dynamic analysis.
  - Because of the expense of equipment. The major concern is to keep such equipment anchored (it is commonly modeled as an inverted pendulum with appendages) and damped to reduce wild excursions.

REFERENCE FIGURE FOR INSTALLATION DETAILS

- 4.75.

RELATIVE DEGREE OF DAMAGE OF INADEQUATELY PROTECTED EQUIPMENT

- Minor to moderate.

MOST LIKELY TYPE OR CONSEQUENCE OF DAMAGE FOR
INADEQUATELY PROTECTED EQUIPMENT

- Toppled equipment.
- Internal mechanical damage to tracks, and so on.
- Possibly inoperative equipment.
- General cleanup required.

Medical Systems

Gas Cylinders

High-pressure gas cylinders (Figure 3.96) could prove lethal should their valves break upon falling. The lightweight chains commonly used around the gas bottles do not keep them from falling.

EQUIPMENT SEISMIC CATEGORY

- "C" support equipment.

SEISMIC SPECIFICATION

- SDS-2.

SEISMIC QUALIFICATION APPROACH

- Design team judgment.
  - Cylinders must be kept from falling during shaking. Nylon straps or lightweight chains will not keep cylinders standing on end.
- Equivalent static coefficient analysis.
  - Design of anchorage systems.

FIGURE 3.96. Unsecured gas cylinders such as these could be dangerous to both life and property during an earthquake.

REFERENCE FIGURE FOR INSTALLATION DETAILS

● 4.59.

RELATIVE DEGREE OF DAMAGE OF INADEQUATELY PROTECTED EQUIPMENT

● Minor to major.

MOST LIKELY TYPE OR CONSEQUENCE OF DAMAGE FOR INADEQUATELY PROTECTED EQUIPMENT

● Cylinders fall and roll around.
● Although no cases have been reported, there is enough energy available in some fully pressurized bottles to propel them through partition walls should the valve be broken off.
● General cleanup required.

REFERENCE FIGURE FOR EXAMPLE OF DAMAGED EQUIPMENT

● 3.190.

Medical Systems

General Lab Equipment

General laboratory equipment (Figure 3.97) is especially vulnerable. The cleanup can be time consuming and some spills may be potentially toxic.

EQUIPMENT SEISMIC CATEGORY

● "C" support equipment.

SEISMIC SPECIFICATION

● SDS-2.

SEISMIC QUALIFICATION APPROACH

● Equivalent static coefficient analysis.
  • Base anchorage for heavy equipment.
● Design team judgment.
  • Shelf retaining wires, lips, and so on.

REFERENCE FIGURES FOR INSTALLATION DETAILS

● 4.53, 4.54, 4.55, 4.56, 4.57, 4.60, 4.61, 4.62, 4.63, 4.68, 4.69, 4.73.

RELATIVE DEGREE OF DAMAGE OF INADEQUATELY PROTECTED EQUIPMENT

● Minor to moderate.

MOST LIKELY TYPE OR CONSEQUENCE OF DAMAGE FOR
INADEQUATELY PROTECTED EQUIPMENT

● Spilled chemicals.
● Broken glassware.
● Spilled shelves.
● Toppled counter top items.
● General cleanup required.

FIGURE 3.97. Typical laboratory scene.

- 3.170, 3.171.

Medical Systems

Gurney

Gurneys (Figure 3.98) should be restrained when not in use.

EQUIPMENT SEISMIC CATEGORY

- "D" support equipment.

SEISMIC SPECIFICATION

- SDS-2.

SEISMIC QUALIFICATION APPROACH

- Design team judgment.
  - Provide storage space for unused equipment.
- Equivalent static coefficient analysis.
  - Method of anchorage when equipment is unused.

REFERENCE FIGURE FOR INSTALLATION DETAILS

- 4.69.

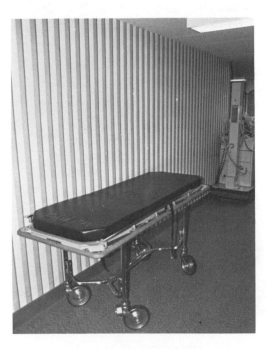

FIGURE 3.98. Unrestrained   gurney stored in corridor.

RELATIVE DEGREE OF DAMAGE OF INADEQUATELY PROTECTED EQUIPMENT

● Minor.

MOST LIKELY TYPE OR CONSEQUENCE OF DAMAGE FOR
INADEQUATELY PROTECTED EQUIPMENT

● Runaway rolling equipment.
● Toppled equipment.
● Collision with other equipment or personnel.
● General cleanup.

Medical Systems

Histology Console

The histology console shown in Figure 3.99 has potential problems similar to those of other consoles. They tend to be top heavy, are seldom anchored, and are subject to operational failures.

EQUIPMENT SEISMIC CATEGORY

● "C" support equipment.

FIGURE 3.99. Unsecured histology console.

SEISMIC SPECIFICATION

● SDS-2.

SEISMIC QUALIFICATION APPROACH

● Equivalent static coefficient analysis.
  • Base anchorage.
● Manufacturer may wish to consider:
  • Dynamic analysis of console frame.
  • Generic test for equipment operability.

REFERENCE FIGURE FOR INSTALLATION DETAILS

● 4.58.

RELATIVE DEGREE OF DAMAGE OF INADEQUATELY PROTECTED EQUIPMENT

● Minor to moderate.

MOST LIKELY TYPE OR CONSEQUENCE OF DAMAGE FOR
INADEQUATELY PROTECTED EQUIPMENT

● Dislodged console.
● Toppled console.
● Potentially inoperative equipment.
● General cleanup required.

Medical Systems

Intensive Care Central Station

Intensive care units are often put on upper floors, which generally experience higher levels of earthquake motions. All intensive care equipment (Figure 3.100) must be given serious aseismic consideration.

FIGURE 3.100. The nursing station cabinet in this example is secured. The monitors, however, are not.

EQUIPMENT SEISMIC CATEGORY

- "A" critical equipment.

SEISMIC SPECIFICATION

- SDS-1.

SEISMIC QUALIFICATION APPROACH

- Equivalent static coefficient analysis.
  - Station console.

REFERENCE FIGURES FOR INSTALLATION DETAILS

- 4.58, 4.60, 4.61, 4.62, 4.63.

RELATIVE DEGREE OF DAMAGE OF INADEQUATELY PROTECTED EQUIPMENT

- Minor to moderate.

MOST LIKELY TYPE OR CONSEQUENCE OF DAMAGE FOR
INADEQUATELY PROTECTED EQUIPMENT

- Movement of console.
- Toppling of console.
- General cleanup required.

Medical Systems

Intensive Care Unit Wall-Mounted Patient Monitors

Wall-mounted patient monitors (Figure 3.101), usually found in Intensive Care Units, must be considered so that they do not fall and injure patients.

EQUIPMENT SEISMIC CATEGORY

- "A" critical equipment.

SEISMIC SPECIFICATION

- SDS-1.

SEISMIC QUALIFICATION APPROACH

- Seismic test.
  - Generic test by manufacturer.
  - Confirms bracket and operability of monitor.
- Dynamic analysis.
  - Confirms bracket design and motions.

REFERENCE FIGURE FOR INSTALLATION DETAILS

- 4.67.

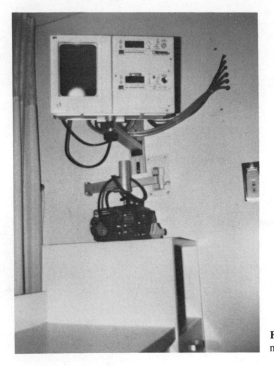

FIGURE 3.101. Wall-mounted patient monitor.

RELATIVE DEGREE OF DAMAGE OF INADEQUATELY PROTECTED EQUIPMENT

● Minor to major.

MOST LIKELY TYPE OR CONSEQUENCE OF DAMAGE FOR INADEQUATELY PROTECTED EQUIPMENT

● Potential for patient injury.
● Potentially inoperable equipment.
● General cleanup required.

REFERENCE FIGURE FOR EXAMPLE OF DAMAGED EQUIPMENT

● 3.172.

Medical Systems

Lighting, Operating Room

Hospital operating room lights (Figure 3.102) are examples of essential lighting. They must remain in place to be functional.

EQUIPMENT SEISMIC CATEGORY

● "A" critical equipment.

FIGURE 3.102. Overhead operating room lights.

SEISMIC SPECIFICATION

- SDS-1.

SEISMIC QUALIFICATION APPROACH

- Equivalent static coefficient analysis.
  - Anchorage.

REFERENCE FIGURE FOR INSTALLATION DETAILS

- 4.64.

RELATIVE DEGREE OF DAMAGE OF INADEQUATELY PROTECTED EQUIPMENT

- Moderate to major.

MOST LIKELY TYPE OR CONSEQUENCE OF DAMAGE FOR
INADEQUATELY PROTECTED EQUIPMENT

- Dislodged lights.
- Potentially inoperative.
- General cleanup required.

Medical Systems

Lighting, X-Ray

X-ray viewing light (see Figure 3.103) is essential for hospital functions after a major earthquake.

EQUIPMENT SEISMIC CATEGORY

- "A" critical equipment.

SEISMIC SPECIFICATION

- SDS-1.

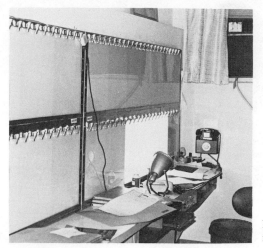

FIGURE 3.103. Unsecured X-ray viewing panels.

SEISMIC QUALIFICATION APPROACH

- Equivalent static coefficient analysis.
  - Anchorage to wall or shelf.

REFERENCE FIGURE FOR INSTALLATION DETAILS

- 4.65.

RELATIVE DEGREE OF DAMAGE OF INADEQUATELY PROTECTED EQUIPMENT

- Minor to moderate.

MOST LIKELY TYPE OR CONSEQUENCE OF DAMAGE FOR
INADEQUATELY PROTECTED EQUIPMENT

- Dislodged bulbs.
- Dislodged/toppled viewing screens.
- Potentially inoperative.
- General cleanup required.

Medical Systems

Liquid Oxygen Supply Tank

Liquid oxygen supply tanks (Figure 3.104) must remain anchored both for hospital function and because of the adverse potential effect of cryogenic spills.

EQUIPMENT SEISMIC CATEGORY

- "A" critical equipment.

FIGURE 3.104. This liquid oxygen tank only received 3 anchor bolts even though the manufacturer provided 12 bolt holes.

SEISMIC SPECIFICATION

- SDS-1.

SEISMIC QUALIFICATION APPROACH

- Equivalent static coefficient analysis.
  - Base anchorage.
- Dynamic analysis.
  - Suggested for consideration of additional sloshing forces by manufacturers to prevent tank rupture.

REFERENCE FIGURES FOR INSTALLATION DETAILS

- 4.66, 4.88.

RELATIVE DEGREE OF DAMAGE OF INADEQUATELY PROTECTED EQUIPMENT

- Moderate to major.

MOST LIKELY TYPE OR CONSEQUENCE OF DAMAGE FOR
INADEQUATELY PROTECTED EQUIPMENT

- Toppled tank.
- Ruptured tank.
- Spilled contents.
- Ruptured supply lines.
- Potential for personnel injury through cryogenic spills.
- General cleanup required.

Medical Systems

Monitors, Operating Room

These monitors (Figure 3.105), whether portable or wall-mounted need close attention with respect to providing earthquake survivability.

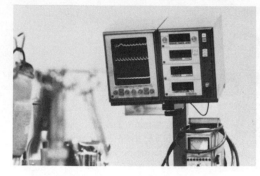

FIGURE 3.105. Example of operating room patient monitor.

EQUIPMENT SEISMIC CATEGORY

- "A" critical equipment.

SEISMIC SPECIFICATION

- SDS-1.

SEISMIC QUALIFICATION APPROACH

- Seismic test.
  - Generic qualification to prove operability.
- Dynamic analysis.
  - Support bracket design.
- Equivalent static coefficient analysis.
  - Bracket base anchorage.
- Design team judgment.
  - Provide restraints for portable equipment.

REFERENCE FIGURES FOR INSTALLATION DETAILS

- 4.67, 4.72.

RELATIVE DEGREE OF DAMAGE OF INADEQUATELY PROTECTED EQUIPMENT

- Minor to major.

MOST LIKELY TYPE OR CONSEQUENCE OF DAMAGE FOR
INADEQUATELY PROTECTED EQUIPMENT

- Toppled equipment.
- Runaway equipment.
- Personnel injury possible.
- Potentially inoperable equipment.
- General cleanup required.

REFERENCE FIGURE FOR EXAMPLE OF DAMAGED EQUIPMENT

● 3.172.

Medical Systems

Monitors, Wheeled CRT

Wheeled CRT monitors (Figure 3.106) are sensitive to earthquake motions and run a high risk of damage if they should fall or collide with other items.

EQUIPMENT SEISMIC CATEGORY

● ''D'' support equipment.

SEISMIC SPECIFICATION

● SDS-2.

SEISMIC QUALIFICATION APPROACH

● Design team judgment.
  • Restrain equipment when not in use.

FIGURE 3.106. Wheel-mounted CRT monitor without provisions for restraining the unit from toppling.

- 4.68, 4.69.

RELATIVE DEGREE OF DAMAGE OF INADEQUATELY PROTECTED EQUIPMENT

- Moderate.

MOST LIKELY TYPE OR CONSEQUENCE OF DAMAGE FOR
INADEQUATELY PROTECTED EQUIPMENT

- Toppled equipment.
- Inoperative equipment.
- Cleanup required.

REFERENCE FIGURE FOR EXAMPLE OF DAMAGED EQUIPMENT

- 3.170.

Medical Systems

Monitor Tree, Intravenous

Intravenous flow monitors (Figure 3.107) tend to roll and fall easily during an earthquake. If they are hooked to a patient when they fall, it could be uncomfortable. Trees supported by the bed itself do not have these problems.

EQUIPMENT SEISMIC CATEGORY

- ''B'' support equipment.

SEISMIC SPECIFICATION

- SDS-1.

SEISMIC QUALIFICATION APPROACH

- Design team judgment.
  - Provide restraint system for equipment both in use and stored.

REFERENCE FIGURE FOR INSTALLATION DETAILS

- 4.68.

RELATIVE DEGREE OF DAMAGE OF INADEQUATELY PROTECTED EQUIPMENT

- Minor to moderate.

MOST LIKELY TYPE OR CONSEQUENCE OF DAMAGE FOR
INADEQUATELY PROTECTED EQUIPMENT

- Toppling of IV tree.
- Potentially inoperative equipment.

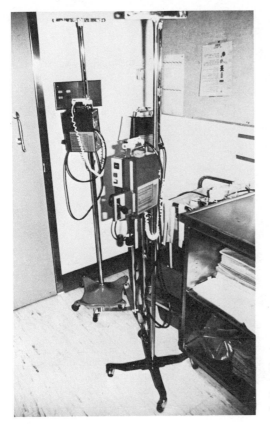

FIGURE 3.107. Tree-mounted intravenous flow monitors. These units are unable to be attached to the patient bed or the wall for stability.

- Potential for personnel injury.
- General cleanup required.

Medical Systems

Operating Room Support Equipment

Specialized operating room equipment (Figure 3.108), much of it portable, must receive qualification consideration so that the operating room can be used immediately after an earthquake as injured patients arrive.

EQUIPMENT SEISMIC CATEGORY

- "A" critical equipment.

SEISMIC SPECIFICATION

- SDS-1.

FIGURE 3.108. Example of operating room support equipment.

SEISMIC QUALIFICATION APPROACH

● Design team judgment.
  • Design for individual pieces of equipment.

REFERENCE FIGURES FOR INSTALLATION DETAILS

● 4.68, 4.69.

RELATIVE DEGREE OF DAMAGE OF INADEQUATELY PROTECTED EQUIPMENT

● Minor to major.

MOST LIKELY TYPE OR CONSEQUENCE OF DAMAGE FOR
INADEQUATELY PROTECTED EQUIPMENT

● Toppled equipment.
● Runaway equipment.
● Personnel injury possible.
● Potentially inoperable equipment.
● General cleanup required.

Medical Systems

Pharmacy Shelves, Rolling

Rolling pharmacy shelves (Figure 3.109) present a dual problem; that of spilled contents and cabinets slamming back and forth.

EQUIPMENT SEISMIC CATEGORY

- "C" support equipment.

SEISMIC SPECIFICATION

- SDS-2.

SEISMIC QUALIFICATION APPROACH

- Design team judgment.
  - Provide retainers for shelved items.
  - Provide guides to keep rolling shelves in their tracks.
  - Keep large bottles and containers on bottom shelves.

REFERENCE FIGURES FOR INSTALLATION DETAILS

- 4.73, 4.74.

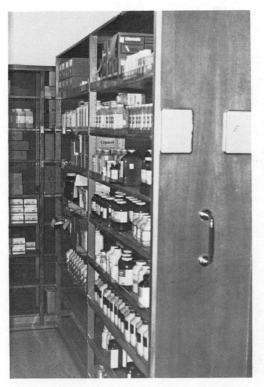

FIGURE 3.109. Rolling pharmacy shelves. Note the Plexiglas shelf parapets.

RELATIVE DEGREE OF DAMAGE OF INADEQUATELY PROTECTED EQUIPMENT

- Minor to moderate.

MOST LIKELY TYPE OR CONSEQUENCE OF DAMAGE FOR
INADEQUATELY PROTECTED EQUIPMENT

- Spilled shelf contents.
- Broken bottles.
- Potential for toxic spills.
- Units coming out of their tracks.
- General cleanup required.

Medical Systems

Pharmacy Work Stations

Pharmacy work spaces (Figure 3.110) generally experience spillage of shelf contents during an earthquake.

EQUIPMENT SEISMIC CATEGORY

- "C" support equipment.

FIGURE 3.110. A typical pharmacy work space. Note the lack of shelf parapets.

SEISMIC SPECIFICATION

- SDS-2.

SEISMIC QUALIFICATION APPROACH

- Design team judgment.
  - Provide shelved item retainers.
  - Provide retainers for counter top equipment such as scales and CRT.
- Equivalent static coefficient analysis.
  - Base anchorage of work stations.

REFERENCE FIGURES FOR INSTALLATION DETAILS

- 4.53, 4.54, 4.55, 4.56, 4.57, 4.60, 4.61, 4.62, 4.63, 4.73, 4.76.

RELATIVE DEGREE OF DAMAGE OF INADEQUATELY PROTECTED EQUIPMENT

- Minor to moderate.

MOST LIKELY TYPE OR CONSEQUENCE OF DAMAGE FOR
INADEQUATELY PROTECTED EQUIPMENT

- Spilled shelf contents.
- Broken glass.
- Potential for toxic spills.
- Dislodged counter top items.
- Displaced work counters.

Medical Systems

Radiological Treatment

Tall, slender, and eccentrically loaded, this equipment is subject to violent motions during an earthquake if it is not adequately supported (Figure 3.111).

EQUIPMENT SEISMIC CATEGORY

- "C" support equipment.

SEISMIC SPECIFICATION

- SDS-2.

SEISMIC QUALIFICATION APPROACH

- Dynamic analysis.
  - For complex equipment such as that shown above.
- Equivalent static coefficient analysis.
  - For less radical equipment configurations.

FIGURE 3.111. Large eccentrically loaded cobalt treatment equipment. This unit is base anchored only.

REFERENCE FIGURE FOR INSTALLATION DETAILS

● 4.75.

RELATIVE DEGREE OF DAMAGE OF INADEQUATELY PROTECTED EQUIPMENT

● Minor to moderate.

MOST LIKELY TYPE OR CONSEQUENCE OF DAMAGE FOR
INADEQUATELY PROTECTED EQUIPMENT

● Equipment malfunction.
● Pounding of adjacent equipment.
● Toppling due to inadequate anchorage design.
● General cleanup required.

Medical Systems

Radiological Treatment Console

Radiological consoles (Figure 3.112) should be treated similarly to communications consoles. Although not critical equipment, their great expense and difficulty to repair warrants a good seismic qualification program.

EQUIPMENT SEISMIC CATEGORY

● "C" support equipment.

FIGURE 3.112. Unsecured radiological treatment console.

SEISMIC SPECIFICATION

- SDS-2 or SDS-1.

SEISMIC QUALIFICATION APPROACH

- Equivalent static coefficient analysis.
  - SDS-2 for base anchorage.
- Dynamic analysis.
  - SDS-1 for the console subcomponents.
- Seismic test.
  - Manufacturers may wish to consider a generic test program for components sensitive to dynamic motions.

REFERENCE FIGURE FOR INSTALLATION DETAILS

- 4.58.

RELATIVE DEGREE OF DAMAGE OF INADEQUATELY PROTECTED EQUIPMENT

- Minor to major.

MOST LIKELY TYPE OR CONSEQUENCE OF DAMAGE FOR
INADEQUATELY PROTECTED EQUIPMENT

- Console displaced.
- Internal component failure.
- Console frame failure.
- Inoperable equipment.
- General cleanup required.

Medical Systems

Respirators

Portable respirators (Figure 3.113) are not necessary for life support but do warrant seismic consideration for reasons such as expense, potential personnel injury, and unnecessary cleanup.

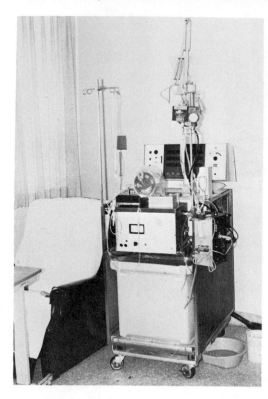

FIGURE 3.113. Portable respirator with wheel locks but no other provision for restraint.

EQUIPMENT SEISMIC CATEGORY

- "D" support equipment.

SEISMIC SPECIFICATION

- SDS-2.

SEISMIC QUALIFICATION APPROACH

- Design team judgment.
  - Provide storage space when equipment is not in use.
- Equivalent static coefficient analysis.
  - Of restraints when equipment is stored.

REFERENCE FIGURES FOR INSTALLATION DETAILS

- 4.68, 4.69.

RELATIVE DEGREE OF DAMAGE OF INADEQUATELY PROTECTED EQUIPMENT

- Minor to moderate.

MOST LIKELY TYPE OR CONSEQUENCE OF DAMAGE FOR
INADEQUATELY PROTECTED EQUIPMENT

- Toppling.
- Collision when set in motion.
- Possibly inoperative.
- General cleanup required.

Medical Systems

Sterile Infant Beds

Provision for anchoring portable infant beds (Figure 3.114) must be given both when the bed is in use and when it is stored.

EQUIPMENT SEISMIC CATEGORY

- "D" support equipment.

SEISMIC SPECIFICATION

- SDS-2.

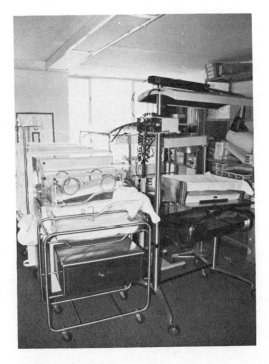

FIGURE 3.114. Stored infant beds without restraint provisions.

SEISMIC QUALIFICATION APPROACH

- Design team judgment.
- Equivalent static coefficient analysis.
  - For anchorage mechanism.

REFERENCE FIGURE FOR INSTALLATION DETAILS

- 4.69.

RELATIVE DEGREE OF DAMAGE OF INADEQUATELY PROTECTED EQUIPMENT

- Minor.

MOST LIKELY TYPE OF CONSEQUENCE OF DAMAGE FOR
INADEQUATELY PROTECTED EQUIPMENT

- Toppled bed.
- Possible infant injury.
- Runaway beds.
- Possible collision with other equipment.
- General cleanup required.

Medical Systems

Storage, Ad Hoc

Unattended storage of miscellaneous equipment (Figure 3.115) must be restrained to prevent monstrous cleanup problems.

EQUIPMENT SEISMIC CATEGORY

- ''E'' miscellaneous equipment.

SEISMIC SPECIFICATION

- SDS-2.

SEISMIC QUALIFICATION APPROACH

- Design team judgment.

REFERENCE FIGURES FOR INSTALLATION DETAILS

- 4.53, 4.54, 4.55, 4.56, 4.60, 4.61, 4.62, 4.63, 4.68, 4.69, 4.73, 4.76, 4.102, 4.103, 4.104.

RELATIVE DEGREE OF DAMAGE OF INADEQUATELY PROTECTED EQUIPMENT

- Minor.

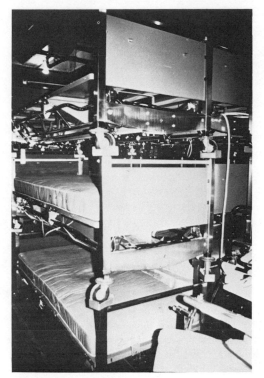

FIGURE 3.115. Ad hoc storage of unrestrained patient beds.

**MOST LIKELY TYPE OR CONSEQUENCE OF DAMAGE FOR
INADEQUATELY PROTECTED EQUIPMENT**

- Toppled stored items.
- General cleanup required.

REFERENCE FIGURE FOR EXAMPLE OF DAMAGED EQUIPMENT

- 3.198.

Medical Systems

Storage, General Shelves

Storage shelves should always be anchored and should always be equipped with shelf restrainers (Figure 3.116).

EQUIPMENT SEISMIC CATEGORY

- "C" support equipment.

SEISMIC SPECIFICATION

- SDS-2.

FIGURE 3.116. Unsecured portable storage shelves without shelf restraints.

- Equivalent static coefficient analysis.
  - Base anchorage.
  - Top anchorage or bracing.
- Design team judgment.
  - Keep heavy items on bottom shelves.
  - Provide shelf restraints.

REFERENCE FIGURES FOR INSTALLATION DETAILS

- 4.53, 4.54, 4.55, 4.56, 4.60, 4.61, 4.62, 4.63, 4.68, 4.69, 4.73, 4.76, 4.102, 4.103, 4.104.

RELATIVE DEGREE OF DAMAGE OF INADEQUATELY PROTECTED EQUIPMENT

- Minor to major.

MOST LIKELY TYPE OR CONSEQUENCE OF DAMAGE FOR INADEQUATELY PROTECTED EQUIPMENT

- Toppled shelves.

- Stored items fall off shelves.
- General cleanup required.

REFERENCE FIGURES FOR EXAMPLES OF DAMAGED EQUIPMENT

- 3.191, 3.192, 3.193, 3.197, 3.198.

Medical Systems

Storage, Records, Hard Copy

Hard copy records require considerable storage space and can create quite a mess if they are not retained on their shelves (Figure 3.117).

EQUIPMENT SEISMIC CATEGORY

- "E" miscellaneous equipment.

SEISMIC SPECIFICATION

- SDS-2.

SEISMIC QUALIFICATION APPROACH

- Equivalent static coefficient analysis.

FIGURE 3.117. Records storage shelves without shelf restraints.

- • Base anchorage of shelves.
- • Top bracing and anchorage of shelves.
- Design team judgment.
  - • Provide shelf restrainers.

REFERENCE FIGURES FOR INSTALLATION DETAILS

- 4.53, 4.54, 4.55, 4.56, 4.60, 4.61, 4.62, 4.63, 4.68, 4.69, 4.73, 4.76, 4.102, 4.103, 4.104.

RELATIVE DEGREE OF DAMAGE OF INADEQUATELY PROTECTED EQUIPMENT

- Minor to moderate.

MOST LIKELY TYPE OR CONSEQUENCE OF DAMAGE FOR INADEQUATELY PROTECTED EQUIPMENT

- Toppled shelves.
- Spilled contents.
- General cleanup required.

REFERENCE FIGURES FOR EXAMPLES OF DAMAGED EQUIPMENT

- 3.191, 3.192, 3.193.

Medical Systems

Storage, Shelving, Wall-Mounted

Wall-mounted shelving (Figure 3.118) must be adequately anchored to prevent collapse.

EQUIPMENT SEISMIC CATEGORY

- "E" miscellaneous equipment.

SEISMIC SPECIFICATION

- SDS-2.

FIGURE 3.118. Typical metal wall-mounted shelves.

SEISMIC QUALIFICATION APPROACH

- Equivalent static coefficient analysis.
- Design team judgment.
  - Provide shelf edge restraints.

REFERENCE FIGURES FOR INSTALLATION DETAILS

- 4.53, 4.54, 4.55, 4.56, 4.73, 4.102, 4.103.

RELATIVE DEGREE OF DAMAGE OF INADEQUATELY PROTECTED EQUIPMENT

- Minor to moderate.

MOST LIKELY TYPE OR CONSEQUENCE OF DAMAGE FOR
INADEQUATELY PROTECTED EQUIPMENT

- If connections are underdesigned, potential collapse of shelving.
- Toppled contents.
- General cleanup required.

Medical Systems

Television, Patient

Patient televisions must remain anchored to prevent injury (Figure 3.119).

EQUIPMENT SEISMIC CATEGORY

- "E" miscellaneous equipment.

SEISMIC SPECIFICATION

- SDS-2.

SEISMIC QUALIFICATION APPROACH

- Equivalent static coefficient analysis.

FIGURE 3.119. Wall-mounted patient television.

- • At a minimum, base anchorage must be considered.
- ● Dynamic analysis.
  - • Manufacturers may wish to generically qualify their supports to assure that the cantilevers are not underdesigned.

REFERENCE FIGURES FOR INSTALLATION DETAILS

- ● 4.67, 4.102, 4.103.

RELATIVE DEGREE OF DAMAGE OF INADEQUATE PROTECTED EQUIPMENT

- ● Minor to moderate.

MOST LIKELY TYPE OR CONSEQUENCE OF DAMAGE FOR INADEQUATELY PROTECTED EQUIPMENT

- ● Fallen televisions.
- ● Potential for personnel injury.
- ● Inoperative televisions.
- ● General cleanup required.

REFERENCE FIGURE FOR EXAMPLE OF DAMAGED EQUIPMENT

- ● 3.172.

Medical Systems

Utensils, Operating Room

Operating room utensils (Figure 3.120) must remain on their storage racks so that they are immediately available for use after an earthquake when injured patients arrive.

EQUIPMENT SEISMIC CATEGORY

- ● "A" critical equipment.

SEISMIC SPECIFICATION

- ● SDS-1.

FIGURE 3.120. Wall-hung operating utensils.

SEISMIC QUALIFICATION APPROACH

- Design team judgment.
  - Use holders that will not allow utensils to fall.

REFERENCE FIGURE FOR INSTALLATION DETAILS

- 4.76.

RELATIVE DEGREE OF DAMAGE OF INADEQUATELY PROTECTED EQUIPMENT

- Minor.

MOST LIKELY TYPE OR CONSEQUENCE OF DAMAGE FOR
INADEQUATELY PROTECTED EQUIPMENT

- Resterilization required if utensils fall to the ground.
- General cleanup required.

Medical Systems

Water Deionizers for Kidney Dialysis Machines

Deionizers for kidney dialysis machines (Figure 3.121) are an example of the specialized application of water systems. Inadequate protection of this

FIGURE 3.121. Unsecured kidney dialysis deionizers.

equipment can be a health threat to individuals undergoing treatment at the time of the earthquake.

EQUIPMENT SEISMIC CATEGORY

- "A" critical equipment.

SEISMIC SPECIFICATION

- SDS-1.

SEISMIC QUALIFICATION APPROACH

- Design team judgment.
  - Provide flexible connections.
- Equivalent static coefficient analysis.
  - Anchorage of filter bottles, pumps, and so on.

REFERENCE FIGURE FOR INSTALLATION DETAILS

- 4.77.

RELATIVE DEGREE OF DAMAGE OF INADEQUATELY PROTECTED EQUIPMENT

- Minor to major.

MOST LIKELY TYPE OR CONSEQUENCE OF DAMAGE FOR
INADEQUATELY PROTECTED EQUIPMENT

- Toppled filter tanks.
- Broken water lines.
- Equipment malfunction.
- Patient injury due to equipment malfunction.
- General cleanup required.

Medical Systems

Wheeled Monitors, Equipment Trees, and Similar Items

Unless restrained, wheeled monitors (Figure 3.122) are likely to topple in an earthquake.

EQUIPMENT SEISMIC CATEGORY

- "D" support equipment.

SEISMIC SPECIFICATION

- SDS-2.

SEISMIC QUALIFICATION APPROACH

- Design team judgment.
  - Design restraint system for unattended equipment.

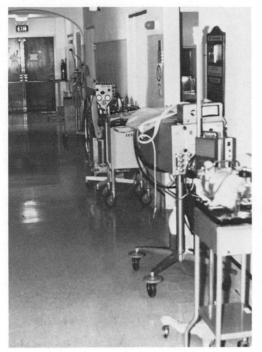

FIGURE 3.122. Wheeled monitors such as those shown here present a serious cleanup problem since this equipment is commonly left unattended in critical hallways. Some of the equipment is likely to be inoperable after toppling.

REFERENCE FIGURES FOR INSTALLATION DETAILS

- 4.68, 4.69.

RELATIVE DEGREE OF DAMAGE OF INADEQUATELY PROTECTED EQUIPMENT

- Minor to moderate.

**MOST LIKELY TYPE OR CONSEQUENCE OF DAMAGE FOR
INADEQUATELY PROTECTED EQUIPMENT**

- Toppled equipment.
- Potentially inoperative equipment.
- General cleanup required.

REFERENCE FIGURE FOR EXAMPLE OF DAMAGED EQUIPMENT

- 3.170.

Medical Systems

X-Ray, Ceiling Track Mounted

Sheet metal tracks for ceiling-mounted X-ray machines (Figure 3.123) are generally not designed to resist earthquake deflections. This problem would be best resolved by manufacturers.

FIGURE 3.123. Track-mounted X-ray camera.

EQUIPMENT SEISMIC CATEGORY

- "A" critical equipment.

SEISMIC SPECIFICATION

- SDS-1.

SEISMIC QUALIFICATION APPROACH

- Dynamic analysis.
  - By manufacturer for equipment design.
- Seismic test.
  - By manufacturer for generic qualification.
- Equivalent static coefficient analysis.
  - By design team for installation if this type of equipment must be used.

REFERENCE FIGURE FOR INSTALLATION DETAILS

- 4.64.

RELATIVE DEGREE OF DAMAGE OF INADEQUATELY PROTECTED EQUIPMENT

- Moderate to major.

MOST LIKELY TYPE OR CONSEQUENCE OF DAMAGE FOR
INADEQUATELY PROTECTED EQUIPMENT

- Dislodged X-ray head unit.
- Inoperable equipment.
- Potential for personnel injury.
- General cleanup required.

Medical Systems

X-Ray, Dental Wall-Mounted Unit

Dental X-ray units (Figure 3.124) may swing wildly during an earthquake with the possibility of striking other equipment or personnel.

EQUIPMENT SEISMIC CATEGORY

- "C" support equipment.

SEISMIC SPECIFICATION

- SDS-2.

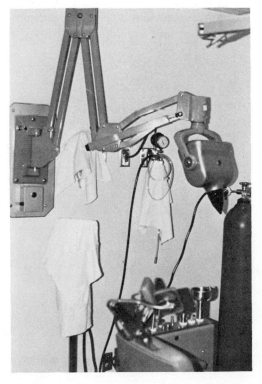

FIGURE 3.124. Wall-mounted dental X-ray unit.

- Design team judgment.
  - Swing arm restrainers when not in use.
- Equivalent static coefficient analysis.
  - Base anchorage at wall connection.

REFERENCE FIGURE FOR INSTALLATION DETAILS

- 4.52.

RELATIVE DEGREE OF DAMAGE FOR INADEQUATELY PROTECTED EQUIPMENT

- None to minor.

MOST LIKELY TYPE OR CONSEQUENCE OF DAMAGE FOR
INADEQUATELY PROTECTED EQUIPMENT

- Possible collision with other equipment while X-ray unit is swinging.

Medical Systems

X-Ray, Fixed

Fixed X-ray equipment is required (Figure 3.125) for adequate patient care immediately after a major earthquake. Equipment subcomponents are commonly precariously or eccentrically mounted and are highly sensitive to dynamic motions.

EQUIPMENT SEISMIC CATEGORY

- "A" critical equipment.

SEISMIC SPECIFICATION

- SDS-1.

SEISMIC QUALIFICATION APPROACH

- Equivalent static coefficient analysis.
  - Base analysis.
- Dynamic analysis.
  - Subcomponents.
  - Frame analysis.
- Seismic test.
  - Generic qualification by manufacturer.

REFERENCE FIGURE FOR INSTALLATION DETAILS

- 4.75.

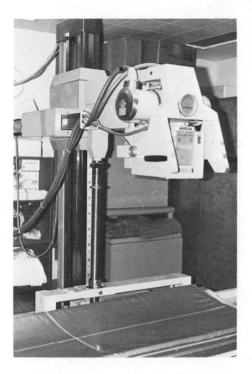

FIGURE 3.125. Typical X-ray units require base anchorage.

RELATIVE DEGREE OF DAMAGE OF INADEQUATELY PROTECTED EQUIPMENT

● Minor to moderate.

MOST LIKELY TYPE OR CONSEQUENCE OF DAMAGE FOR INADEQUATELY PROTECTED EQUIPMENT

● Equipment malfunction.
● Frame racking.
● Equipment toppling because of underdesigned anchorage.
● Equipment pounding.
● General cleanup required.

Medical Systems

X-Ray, Portable

Portable X-ray equipment (Figure 3.126) is commonly left unanchored in hallways and work spaces when it is not in use, inviting disaster.

EQUIPMENT SEISMIC CATEGORY

● "A" critical equipment.

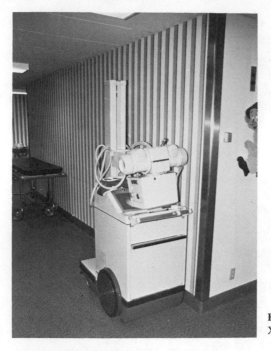

FIGURE 3.126. Unsecured portable X-ray unit.

SEISMIC SPECIFICATION

- SDS-1.

SEISMIC QUALIFICATION APPROACH

- Design team judgment.
  - Provide storage space when equipment is not in use.
- Equivalent static coefficient approach.
  - Anchorage when equipment is not in use.

REFERENCE FIGURE FOR INSTALLATION DETAILS

- 4.69.

RELATIVE DEGREE OF DAMAGE OF INADEQUATELY PROTECTED EQUIPMENT

- Minor to moderate.

MOST LIKELY TYPE OR CONSEQUENCE OF DAMAGE
FOR INADEQUATELY PROTECTED EQUIPMENT

- Toppling.
- Runaway equipment.
- Collision with other equipment.
- Equipment malfunction.
- General cleanup required.

Motion Restraint Systems (Courtesy California Dynamics Corporation)

Unless protected, resiliently supported equipment is vulnerable to earthquake caused damage. Excessive displacements relative to the building can tear connections and excessive equipment velocities endanger equipment and surroundings from hammering impacts. Motion restraints, if properly employed by the design team, can protect the equipment without compromising day-to-day vibration isolation performance unless earthquake criteria are unusually severe.

A motion restraint (or snubber) must be structurally adequate and properly anchored or it will merely give the illusion of protection. The low tensile strength of concrete often dictates extraordinary measures such as embedment of steel beams in concrete floors. Proper design, however, can often avoid this costly procedure and drilled in anchor bolts can be used instead. Proper anchorage design is critical to the successful employment of any motion restraining device. Use of an Integral Vibration Isolation and Snubbing Device (ISOLATOR RESTRAINT) simplifies installation with a minimum quantity of devices, places snubbing loads at the most desirable locations, and reduces pullout loads on anchors by utilizing the equipment weight for minimal anchorage difficulties.

The design team will generally find the best route to qualification by having the motion restraint system designed by the professional design staff of the manufacturer. The individual motion restraint manufacturers are equipped to solve the installation problem with minimal input from the design team. Required information includes:

- Equipment data and geometry.
- Floor motion criteria.
- Response spectrums (or other dynamic criteria).
- Applicable code requirements.

SYSTEM SEISMIC CATEGORY

- "A" critical system.

SYSTEM FOUND IN

- All facilities with resiliently supported equipment.

Motion Restraint Systems

All-Directional Snubbers

All-directional snubbers are generally cylindrical and each snubber provides for equipment motion in all three orthogonal axes. They can be installed alongside or beneath equipment. All directional snubbers are generally installed on two sides of the reciprocating equipment adjacent to the vibration isolators.

EQUIPMENT SEISMIC CATEGORY

- "A" critical equipment.

SEISMIC SPECIFICATION

- SDS-1.

SEISMIC QUALIFICATION APPROACH

- Design team judgment.
  - Select motion restraint manufacturer.
  - Refer to professional motion restraint manufacturer for selection of suitable restraint type and installation specifications.

REFERENCE FIGURES FOR INSTALLATION DETAILS

- 4.78, 4.79, 4.80, 4.81, 4.86.

RELATIVE DEGREE OF DAMAGE FOR INADEQUATELY PROTECTED EQUIPMENT

- Moderate to major.

MOST LIKELY TYPE OR CONSEQUENCE OF DAMAGE FOR
INADEQUATELY PROTECTED EQUIPMENT

- Equipment dislocation.
- Severed supply lines.
- Equipment failure.

Motion Restraint Systems

Isolator Restraints

Isolator restraints are integral vibration isolators and motion restraints that have been combined into a single package. These isolators are easy to examine and maintain.

EQUIPMENT SEISMIC CATEGORY

- "A" critical equipment.

SEISMIC SPECIFICATION

- SDS-1.

SEISMIC QUALIFICATION APPROACH

- Design team judgment.
  - Select motion restraint manufacturer.
  - Refer to professional motion restraint manufacturer for selection of suitable restraint type and installation specifications.

- 4.78, 4.79, 4.80, 4.81, 4.83, 4.84, 4.85.

RELATIVE DEGREE OF DAMAGE FOR INADEQUATELY PROTECTED EQUIPMENT

- Moderate to major.

MOST LIKELY TYPE OR CONSEQUENCE OF DAMAGE FOR
INADEQUATELY PROTECTED EQUIPMENT

- Equipment dislocation.
- Severed supply lines.
- Equipment failure.

Motion Restraint Systems

Lockout Devices

Lockout devices are generally pneumatically operated and restrain equipment that has been resiliently mounted from undergoing sympathetic vibrations that may result from equipment/building–isolation interactions. When a seismic sensing device on the motion restraint "feels" building motions greater than anticipated under normal operation, the lockout is employed. Rams that prevent further motion of the equipment are then inserted into receiving cups on the equipment skid. Once tripped, lockout devices must be manually reset.

EQUIPMENT SEISMIC CATEGORY

- "A" critical equipment.

SEISMIC SPECIFICATION

- SDS-1.

SEISMIC QUALIFICATION APPROACH

- Design team judgment.
  - Select motion restraint manufacturer.
  - Refer to professional motion restraint manufacturer for selection of suitable restraint type and installation specifications.

REFERENCE FIGURES FOR INSTALLATION DETAILS

- 4.39, 4.78, 4.79, 4.80, 4.81.

RELATIVE DEGREE OF DAMAGE FOR INADEQUATELY PROTECTED EQUIPMENT

- Moderate to major.

MOST LIKELY TYPE OR CONSEQUENCE OF DAMAGE FOR
INADEQUATELY PROTECTED EQUIPMENT

- Equipment dislocation.
- Severed supply lines.
- Equipment failure.

Motion Restraint Systems

Snubbers, Angle Stops

For selected pieces of noncritical and inexpensive reciprocating equipment, the design team may wish to design its own snubbers. The most common method uses angle stops with rubber inertia pads. Care must be taken to avoid creating shock loads that can sever bolt heads, damage the equipment itself, or break the bond between the concrete and anchor bolts. Anchor bolt holes in the base must be round and generally the same size as the anchor bolt rather than oblong.

EQUIPMENT SEISMIC CATEGORY

- "B" support equipment.

SEISMIC SPECIFICATION

- SDS-1.

SEISMIC QUALIFICATION APPROACH

- Design team judgment.
  - Make decision for in-house design of snubbers.
- Dynamic analysis.
  - Determine shock loads.
  - Determine spring-mass response (equipment displacements, etc.).
  - Determine anchorage characteristics.
  - Determine correct tolerances.

REFERENCE FIGURES FOR INSTALLATION DETAILS

- 4.78, 4.79, 4.80, 4.81, 4.82.

RELATIVE DEGREE OF DAMAGE FOR INADEQUATELY PROTECTED EQUIPMENT

- Moderate to major.

MOST LIKELY TYPE OR CONSEQUENCE OF DAMAGE FOR
INADEQUATELY PROTECTED EQUIPMENT

- Equipment dislocation.
- Severed supply lines.
- Equipment failure.

Piping Systems

Although much of the piping within a facility is not critical in itself, ruptures at an inappropriate place and time can lead to the failure of other critical equipment items; for example, a water line failure that allows the emergency power supply room to flood, thus shutting down facility power. The basic philosophy behind qualification of piping systems is to keep the line suspended and to keep it from deflecting so much that it is likely to rupture. Bracing and flexible connections along with separation of the pipe run from the building structure/pipe interface (i.e., where a pipe passes through a wall) significantly reduce piping system failures. Obviously, for critical supply lines or where a failure of a noncritical line is likely to affect the operation of an adjacent piece of equipment in a higher seismic category, the pipe run must be "tuned" so that its natural frequency does not fall within the frequency generated by the earthquake.

SYSTEM SEISMIC CATEGORY

● "B" support system.

SYSTEM FOUND IN

● All facilities.

Piping Systems

Pipe at Seismic Joints

Piping at seismic joints must be capable of displacement in three orthogonal axes simultaneously through the use of flexible connections or ball joints. The inability of the pipe to move with the building would otherwise result in a likelihood of pipe failure.

EQUIPMENT SEISMIC CATEGORY

● Varies.

SEISMIC SPECIFICATION

● SDS-1 or SDS-2.

SEISMIC QUALIFICATION APPROACH

● Design team judgment.
● Dynamic analysis.
● Pipe flex computer programs are available.

REFERENCE FIGURE FOR INSTALLATION DETAILS

● 4.93.

RELATIVE DEGREE OF DAMAGE FOR INADEQUATELY PROTECTED EQUIPMENT

- Minor to major.

MOST LIKELY TYPE OR CONSEQUENCE OF DAMAGE FOR
INADEQUATELY PROTECTED EQUIPMENT

- Severed pipe.
- Flooding.
- Sanitation problems.
- Steam escape.

Piping Systems

Pipe Hangers, Lateral Braced Horizontal Pipe

Unbraced horizontal pipe is subject to pipe sway, which can transmit high loads to fixed flanges. Lateral bracing (Figure 3.127) can reduce the potential for this type of failure. Compression posts prevent vertical pipe motions.

EQUIPMENT SEISMIC CATEGORY

- Varies.

SEISMIC SPECIFICATION

- SDS-1 or SDS-2.

SEISMIC QUALIFICATION APPROACH

- Design team judgment.
- Dynamic analysis.
- Pipe flex computer programs are available.

REFERENCE FIGURES FOR INSTALLATION DETAILS

- 4.89, 4.95.

RELATIVE DEGREE OF DAMAGE FOR INADEQUATELY PROTECTED EQUIPMENT

- Minor to major.

MOST LIKELY TYPE OR CONSEQUENCE OF DAMAGE FOR
INADEQUATELY PROTECTED EQUIPMENT

- Severed pipe.
- Flooding.
- Sanitation problems.
- Steam escape.

REFERENCE FIGURES FOR EXAMPLES OF DAMAGED EQUIPMENT

- 3.182, 3.183.

FIGURE 3.127. Horizontal pipe with hangers and lateral bracing.

Piping Systems

Pipe Hangers, Longitudinal Braced Horizontal Pipe

Pipe that is unbraced in the longitudinal direction is subject to axial loads at fixed points that may lead to ruptures. Compression posts prevent vertical pipe motions.

EQUIPMENT SEISMIC CATEGORY

- Varies.

SEISMIC SPECIFICATION

- SDS-1 or SDS-2.

SEISMIC QUALIFICATION APPROACH

- Design team judgment.
- Dynamic analysis.
- Pipe flex computer programs are available.

REFERENCE FIGURE FOR INSTALLATION DETAILS

- 4.90.

- Minor to major.

MOST LIKELY TYPE OR CONSEQUENCE OF DAMAGE FOR
INADEQUATELY PROTECTED EQUIPMENT

- Severed pipe.
- Flooding.
- Sanitation problems.
- Steam escape.

REFERENCE FIGURES FOR EXAMPLES OF DAMAGED EQUIPMENT

- 3.182, 3.183.

Piping Systems

Pipe Hangers, Vertical Pipe

Even where the pipe hanger (Figure 3.128) is set in the floor slab, the pipe itself must not be rigidly attached to the slab unless flexible connectors are supplied on either side of the slab.

EQUIPMENT SEISMIC CATEGORY

- Varies—Rupture can often affect critical equipment.

SEISMIC SPECIFICATION

- SDS-1 or SDS-2.

SEISMIC QUALIFICATION APPROACH

- Design team judgment.
- Dynamic analysis.
- Pipe flex computer programs are available.

FIGURE 3.128. Vertical pipe hanger at floor level. Note the waste line bisecting the partition wall.

- 4.91.

RELATIVE DEGREE OF DAMAGE FOR INADEQUATELY PROTECTED EQUIPMENT

- Minor to major.

MOST LIKELY TYPE OR CONSEQUENCE OF DAMAGE FOR
INADEQUATELY PROTECTED EQUIPMENT

- Severed pipe.
- Flooding.
- Sanitation problems.
- Steam escape.

Piping Systems

Pipe Racks

Pipe racks can be designed to allow for two or three controlled directions of motion. Shock loads must be considered at all pipe stops.

EQUIPMENT SEISMIC CATEGORY

- Varies.

SEISMIC SPECIFICATION

- SDS-1 or SDS-2.

SEISMIC QUALIFICATION APPROACH

- Design team judgment.
- Dynamic analysis.
- Pipe flex computer programs are available.

RELATIVE DEGREE OF DAMAGE FOR INADEQUATELY PROTECTED EQUIPMENT

- Minor to major.

MOST LIKELY TYPE OR CONSEQUENCE OF DAMAGE FOR
INADEQUATELY PROTECTED EQUIPMENT

- Severed pipe.
- Flooding.
- Sanitation problems.
- Steam escape.

REFERENCE FIGURES FOR EXAMPLES OF DAMAGED EQUIPMENT

- 3.180, 3.181.

Piping Systems

Tubing/Conduit

Tubing must be restrained for exceptionally long runs. It generally fairs well if flexible connections are provided at wall intersections, machinery interfaces, and so on (Figure 3.129).

EQUIPMENT SEISMIC CATEGORY

- Varies.

SEISMIC SPECIFICATION

- SDS-1 or SDS-2.

SEISMIC QUALIFICATION APPROACH

- Design team judgment.
  - Provide flexible connectors.
  - Provide support for long runs.

REFERENCE FIGURES FOR INSTALLATION DETAILS

- 4.87, 4.88, 4.92.

RELATIVE DEGREE OF DAMAGE FOR INADEQUATELY PROTECTED EQUIPMENT

- Minor to major.

MOST LIKELY TYPE OR CONSEQUENCE OF DAMAGE FOR
INADEQUATELY PROTECTED EQUIPMENT

- Severed pipe.
- Flooding.
- Sanitation problems.
- Steam escape.

FIGURE 3.129.   Flexible tubing at floor allows for movement without rupture.

Suspended Ceiling Systems

Suspended ceiling systems are not required for facility operation. Their failure does, however, present efficiency problems immediately after an earthquake. Those within a facility at the time of an earthquake may suffer undue "psychological stress" if ceiling panels begin to fall.

SYSTEM SEISMIC CATEGORY

- "E" miscellaneous system.

SYSTEM FOUND IN

- Business establishments.
- Communication centers.
- Computing/data processing centers.
- Emergency operating centers.
- Fire stations.
- Government administration buildings.
- Hospitals.
- Police stations.
- Schools.

Suspended Ceiling Systems

T-Bar Ceiling

T-bar ceilings (Figure 3.130) are highly susceptible to earthquakes. Lay-in panels commonly fall as the T-bar frame deflects.

EQUIPMENT SEISMIC CATEGORY

- "E" miscellaneous equipment.

FIGURE 3.130. T-bar ceilings with lay-in panels. Photograph courtesy of Ruhnau · Evans · Ruhnau · Associates.

SEISMIC SPECIFICATION

- SDS-2.

SEISMIC QUALIFICATION APPROACH

- Equivalent static coefficient analysis.
  - T-bar frame.
- Design team judgment.
  - Specify compression posts and diagonal wiring.

REFERENCE FIGURE FOR INSTALLATION DETAILS

- 4.98.

RELATIVE DEGREE OF DAMAGE OF INADEQUATELY PROTECTED EQUIPMENT

- Minor.

MOST LIKELY TYPE OR CONSEQUENCE OF DAMAGE FOR
INADEQUATELY PROTECTED EQUIPMENT

- Dislodged lay-in panels.
- Cleanup of collapsed panels required.
- Facility personnel unrest as panels fall. They may believe that the building is falling and panic.

REFERENCE FIGURE FOR EXAMPLE OF DAMAGED EQUIPMENT

- 3.185.

Water Systems

Water systems include more than an occasional water pipe or sink. General performance in past earthquakes has been good. Qualification must be considered, however, to prevent pipe ruptures and subsequent flooding, which could prevent the operation of other systems.

SYSTEM SEISMIC CATEGORY

- ''B'' support equipment.

SYSTEM FOUND IN

- All building types.

Water Systems

Boilers

Boilers for facility heating are typically large and earthquake consideration

FIGURE 3.131. Large unsecured boiler.

is generally not given (Figure 3.131). This equipment must receive adequate protection to remain operational.

EQUIPMENT SEISMIC CATEGORY

● ''B'' support equipment.

SEISMIC SPECIFICATION

● SDS-1.

SEISMIC QUALIFICATION APPROACH

● Equivalent static coefficient analysis.
  • Anchorage.
● Design team judgment.
  • Use flexible connections on supply lines (water and gas)

REFERENCE FIGURE FOR INSTALLATION DETAILS

● 4.100.

RELATIVE DEGREE OF DAMAGE OF INADEQUATELY PROTECTED EQUIPMENT

● Minor to moderate.

- Dislodged boiler.
  - Excessive movement possible if unanchored.
- Possibly inoperative boiler.
- Ruptured supply lines.
  - Flooding potential.
  - Fire potential.
- General cleanup required.

Water Systems

Drinking Fountains, Freestanding

Freestanding water coolers (Figure 3.132) are the type most likely to cause damage (flooding) as a result of an earthquake.

EQUIPMENT SEISMIC CATEGORY

- "E" miscellaneous equipment.

SEISMIC SPECIFICATION

- SDS-2.

SEISMIC QUALIFICATION APPROACH

- Equivalent static coefficient analysis.
  - Base anchorage.
- Design team judgment.
  - Provide flexible water lines to prevent rupture.
  - Use top bracing if possible.

REFERENCE FIGURE FOR INSTALLATION DETAILS

- 4.35.

RELATIVE DEGREE OF DAMAGE OF INADEQUATELY PROTECTED EQUIPMENT

- Minor to moderate.

MOST LIKELY TYPE OR CONSEQUENCE OF DAMAGE FOR
INADEQUATELY PROTECTED EQUIPMENT

- Toppled drinking fountain.
- Broken water lines (flooding).
- General cleanup required.

FIGURE 3.132. Freestanding water cooler. Photograph courtesy of Oasis Water Coolers.

Water Systems

Drinking Fountains, Wall-Hung

Wall-hung drinking fountains (Figure 3.133) have good performance records. Anchorage must be adequate especially for cantilevered models.

EQUIPMENT SEISMIC CATEGORY

- ''E'' miscellaneous equipment.

SEISMIC SPECIFICATION

- SDS-2.

FIGURE 3.133. Wall-hung water cooler. Photograph courtesy of Oasis Water Coolers.

SEISMIC QUALIFICATION APPROACH

- Equivalent static coefficient analysis.
  - For anchorage.
- Design team judgment.
  - Provide flexible supply line connection.

REFERENCE FIGURES FOR INSTALLATION DETAILS

- 4.102, 4.103.

RELATIVE DEGREE OF DAMAGE OF INADEQUATELY PROTECTED EQUIPMENT

- Minor.

MOST LIKELY TYPE OR CONSEQUENCE OF DAMAGE FOR
INADEQUATELY PROTECTED EQUIPMENT

- Dislodged if inadequate anchorage on cantilevered models.
- Ruptured supply lines.
- Flooding potential.
- General cleanup required.

Water Systems

Hot and Cold Water Supply Lines

Most water supply lines (Figure 3.134) fair generally well during an earthquake.

EQUIPMENT SEISMIC CATEGORY

● "B" support equipment.

SEISMIC SPECIFICATION

● SDS-1.

SEISMIC QUALIFICATION APPROACH

● Design team judgment.
  • For small lines.
● Equivalent static coefficient analysis.
  • For larger line supports (2 inches and above) and braces.

REFERENCE FIGURE FOR INSTALLATION DETAILS

● 4.92.

RELATIVE DEGREE OF DAMAGE OF INADEQUATELY PROTECTED EQUIPMENT.

● Minor to moderate.

MOST LIKELY TYPE OR CONSEQUENCE OF DAMAGE FOR
INADEQUATELY PROTECTED EQUIPMENT

● System generally operative. Most frequent types of damage are minor leaks and broken supports.
● Potential for serious flooding should large lines rupture.
● General cleanup required.

FIGURE 3.134. Typical water supply lines. Note the plate welded to the metal studs that will be used for hanging the casework.

Water Systems

Pumps

Water pumps (Figure 3.135) are common victims of earthquake forces, especially when motion restraints are not used in conjunction with vibration isolators and when flexible line connectors are not employed.

EQUIPMENT SEISMIC CATEGORY

- "B" support equipment.

SEISMIC SPECIFICATION

- SDS-1.

SEISMIC QUALIFICATION APPROACH

- Design team judgment.
  - Provide flexible supply lines.
- Equivalent static coefficient analysis.
  - Fixed anchorage.
- Dynamic analysis.
  - Vibration isolation, provide snubbers.

REFERENCE FIGURE FOR INSTALLATION DETAILS

- 4.99.

RELATIVE DEGREE OF DAMAGE OF INADEQUATELY PROTECTED EQUIPMENT

- Minor to moderate.

MOST LIKELY TYPE OR CONSEQUENCE OF DAMAGE FOR INADEQUATELY PROTECTED EQUIPMENT

- Displaced pumps.

FIGURE 3.135. Chilled water pump with flexible connections. The pump is anchored to the inertia pad, which is supported by four unsecured vibration isolators. The lack of motion restraint and the unsecured vibration isolation are likely to result in pump failure during an earthquake.

- Inoperative pumps.
- Severed supply lines.
- Flooding potential.
- Vibration isolation failure.
- General cleanup required.

REFERENCE FIGURE FOR EXAMPLE OF DAMAGED EQUIPMENT

- 3.178.

Water Systems

Storage Tanks

Above ground storage tanks (Figure 3.136) require base anchorage and flexible supply line connections to prevent possible flooding.

EQUIPMENT SEISMIC CATEGORY

- "B" support equipment.

SEISMIC SPECIFICATION

- SDS-1.

FIGURE 3.136. Large, unsecured water storage tank without flexible connections.

SEISMIC QUALIFICATION APPROACH

- Equivalent static coefficient analysis.
  - Tank anchorage.
- Design team judgment.
  - Provide flexible supply line connections.

REFERENCE FIGURES FOR INSTALLATION DETAILS

- 4.100, 4.101.

RELATIVE DEGREE OF DAMAGE OF INADEQUATELY PROTECTED EQUIPMENT

- Minor to major.

MOST LIKELY TYPE OR CONSEQUENCE OF DAMAGE FOR
INADEQUATELY PROTECTED EQUIPMENT

- Excessive tank displacement if unanchored.
- Ruptured supply lines.
- Flooding potential.

REFERENCE FIGURES FOR EXAMPLES OF DAMAGED EQUIPMENT

- 3.186, 3.187, 3.188, 3.189.

Water Systems

Water Heaters

Hot water heaters (Figure 3.137) are usually tall, slender pieces of equipment that are highly susceptible to earthquake motions.

EQUIPMENT SEISMIC CATEGORY

- "B" support equipment.

SEISMIC SPECIFICATION

- SDS-1.

SEISMIC QUALIFICATION APPROACH

- Equivalent static coefficient analysis.
  - Base and top anchorage.
- Design team judgment.
  - Provide flexible connectors.

REFERENCE FIGURE FOR INSTALLATION DETAILS

- 4.101.

RELATIVE DEGREE OF DAMAGE OF INADEQUATELY PROTECTED EQUIPMENT

- Minor to moderate.

FIGURE 3.137. Hot water storage tank without base anchorage or flexible line connectors. Pipe legs are brittle and commonly fail during earthquakes.

MOST LIKELY TYPE OR CONSEQUENCE OF DAMAGE FOR
INADEQUATELY PROTECTED EQUIPMENT

- Broken pipe leg supports.
- Toppled tanks.
- Severed supply lines (gas and water).
  - Flooding potential.
  - Fire potential.
- Secondary damage due to toppled tanks colliding with other equipment.
- Hot water system left inoperative.
- General cleanup required.

REFERENCE FIGURES FOR EXAMPLES OF DAMAGED EQUIPMENT

- 3.186, 3.187, 3.188, 3.189.

Water Systems

Water Softeners

Water softeners (Figure 3.138) commonly have a geometry similar to that of hot water heaters and, consequently, similar stability problems.

FIGURE 3.138.   Water softener.

EQUIPMENT SEISMIC CATEGORY

- "C" support equipment.

SEISMIC SPECIFICATION

- SDS-2.

SEISMIC QUALIFICATION APPROACH

- Equivalent static coefficient analysis.
  - Base and top anchorage.
- Design team judgment.
  - Provide flexible connections.

REFERENCE FIGURE FOR INSTALLATION DETAILS

- 4.101.

RELATIVE DEGREE OF DAMAGE OF INADEQUATELY PROTECTED EQUIPMENT

- Minor to moderate.

**MOST LIKELY TYPE OR CONSEQUENCE OF DAMAGE FOR
INADEQUATELY PROTECTED EQUIPMENT**

- Pipe leg failure.
- Toppled tanks.
- Severed supply lines with flooding potential.
- General cleanup required.

REFERENCE FIGURES FOR EXAMPLES OF DAMAGED EQUIPMENT

- 3.186, 3.187, 3.188, 3.189.

Water Systems

Water Wells

Water wells (Figure 3.139) require protection from other equipment to assure their operability. Collapsed casings and so forth are beyond the scope of this book with respect to well production.

EQUIPMENT SEISMIC CATEGORY

- "B" support equipment.

SEISMIC SPECIFICATION

- SDS-1.

SEISMIC QUALIFICATION APPROACH

- Design team judgment.
  - For surrounding equipment.

RELATIVE DEGREE OF DAMAGE OF INADEQUATELY PROTECTED EQUIPMENT

- Minor to moderate.

FIGURE 3.139. Water well.

MOST LIKELY TYPE OR CONSEQUENCE OF DAMAGE FOR
INADEQUATELY PROTECTED EQUIPMENT

- Damage from unanchored adjacent equipment possible.
- Inoperative well.
- Flooding potential.

Miscellaneous Equipment

The equipment items noted in this section are commonly found in nearly all facilities and do not easily fit into any of the described systems above.

Miscellaneous Equipment

Bookshelves

Bookshelves (Figure 3.140) are quite heavy when fully loaded and can present a cleanup problem if the books themselves are not restrained.

EQUIPMENT SEISMIC CATEGORY

- "E" miscellaneous equipment.

SEISMIC SPECIFICATION

- SDS-2.

SEISMIC QUALIFICATION APPROACH

- Equivalent static coefficient analysis.
  - Base anchorage.
  - Longitudinal "×" bracing.
  - Top bracing.
- Design team judgment.
  - Provide shelf restrainers or tilted shelves.

FIGURE 3.140. These bookshelves have been anchored to the floor and have lateral bracing in the longitudinal direction. They lack top bracing and shelf parapets.

REFERENCE FIGURES FOR INSTALLATION DETAILS

● 4.102, 4.103, 4.104.

RELATIVE DEGREE OF DAMAGE OF INADEQUATELY PROTECTED EQUIPMENT

● Minor.

MOST LIKELY TYPE OR CONSEQUENCE OF DAMAGE FOR
INADEQUATELY PROTECTED EQUIPMENT

● Toppled shelves.
● Major cleanup required.

REFERENCE FIGURE FOR EXAMPLE OF DAMAGED EQUIPMENT

● 3.193.

Miscellaneous Equipment

Counters

Counters need adequate anchorage (Figure 3.141).

EQUIPMENT SEISMIC CATEGORY

● ''E'' miscellaneous equipment.

SEISMIC SPECIFICATION

● SDS-2.

SEISMIC QUALIFICATION APPROACH

● Equivalent static coefficient analysis.
  • Base anchorage.
● Design team judgment.
  • Use positive latch doors.
  • Use drawer stops.

FIGURE 3.141. Anchorage must be provided for all counters whether they are located against a wall or out in the working spaces.

REFERENCE FOR INSTALLATION DETAILS

● Appendix 3.

RELATIVE DEGREE OF DAMAGE OF INADEQUATELY PROTECTED EQUIPMENT

● Minor.

MOST LIKELY TYPE OR CONSEQUENCE OF DAMAGE FOR
INADEQUATELY PROTECTED EQUIPMENT

● Dislodged counters.
● Spilled items.
● General cleanup required.

Miscellaneous Equipment

Desks

Sliding or toppling desks present both a cleanup problem as well as the potential for personnel injury (Figure 3.142).

EQUIPMENT SEISMIC CATEGORY

● "E" miscellaneous equipment.

SEISMIC SPECIFICATION

● SDS-2.

FIGURE 3.142. Office desks unless properly cared for can shift or even topple with the potential for injury. Drawers should be kept closed to reduce the toppling potential.

SEISMIC QUALIFICATION APPROACH

- Equivalent static coefficient analysis.
  - Base anchorage.

REFERENCE FIGURES FOR INSTALLATION DETAILS

- 4.105, 4.106, 4.107.

RELATIVE DEGREE OF DAMAGE OF INADEQUATELY PROTECTED EQUIPMENT

- Minor.

MOST LIKELY TYPE OR CONSEQUENCE OF DAMAGE FOR
INADEQUATELY PROTECTED EQUIPMENT

- Dislodged desks.
- Toppled desks.
- Potential for personnel injury.
- General cleanup required.

REFERENCE FIGURE FOR EXAMPLE OF DAMAGED EQUIPMENT

- 3.197.

Miscellaneous Equipment

Filing Cabinets

Filing cabinets (Figure 3.143) have been known to fall and block major exit ways in numerous facilities. The general cleanup and possibility of personnel injury should also be considered.

EQUIPMENT SEISMIC CATEGORY

- "E" miscellaneous equipment.

SEISMIC SPECIFICATION

- SDS-2.

SEISMIC QUALIFICATION APPROACH

- Equivalent static coefficient analysis.
  - Base anchorage where possible.
- Design team judgment.
  - Provide top clips.
  - Provide top bracing.
  - Keep drawers at the bottom the fullest.
  - Keep drawers closed.
  - Purchase only filing cabinets with drawer locks.

FIGURE 3.143. Filing cabinets if properly detailed, as these are, will not present a problem in the event of an earthquake. Photograph courtesy of Ruhnau · Evans · Ruhnau · Associates.

REFERENCE FIGURES FOR INSTALLATION DETAILS

● 4.108, 4.109.

RELATIVE DEGREE OF DAMAGE OF INADEQUATELY PROTECTED EQUIPMENT

● Minor.

MOST LIKELY TYPE OR CONSEQUENCE OF DAMAGE FOR INADEQUATELY PROTECTED EQUIPMENT

● Toppled filing cabinets.
● General cleanup required.
● Potential for personnel injury.

REFERENCE FIGURES FOR EXAMPLES OF DAMAGED EQUIPMENT

● 3.194, 3.195, 3.196.

Miscellaneous Equipment

Monitors, CRT

CRT monitors (Figure 3.144) should be installed per manufacturer recommendations.

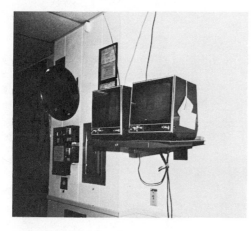

FIGURE 3.144. These unsecured CRT monitors have been installed on a shelf that is designed for one set. Note that they are sitting on a thin board because they were too large for the shelf.

EQUIPMENT SEISMIC CATEGORY

- "C" support equipment.

SEISMIC SPECIFICATION

- SDS-2.

SEISMIC QUALIFICATION APPROACH

- Equivalent static coefficient analysis.
  - Bracket anchorage to wall.
  - CRT anchorage to bracket.
- Dynamic analysis.
  - Manufacturers may wish to perform generic qualification programs for mounting monitors on their brackets.

REFERENCE FIGURE FOR INSTALLATION DETAILS

- 4.67.

RELATIVE DEGREE OF DAMAGE OF INADEQUATELY PROTECTED EQUIPMENT

- Minor to moderate.

MOST LIKELY TYPE OR CONSEQUENCE OF DAMAGE FOR INADEQUATELY PROTECTED EQUIPMENT

- Toppled CRTs.
- Inoperative equipment.
- Potential for personnel injury.
- General cleanup required.

REFERENCE FIGURE FOR EXAMPLE OF DAMAGED EQUIPMENT

- 3.172.

Miscellaneous Equipment

Personnel Lockers

These items need anchorage to prevent toppling (Figure 3.145).

EQUIPMENT SEISMIC CATEGORY

● "E" miscellaneous equipment.

SEISMIC SPECIFICATION

● SDS-2.

SEISMIC QUALIFICATION APPROACH

● Equivalent static coefficient analysis.
  • Base anchorage.
  • Attach to walls where possible.
  • Braced tops.

REFERENCE FIGURES FOR INSTALLATION DETAILS

● 4.108, 4.109.

RELATIVE DEGREE OF DAMAGE OF INADEQUATELY PROTECTED EQUIPMENT

● Minor to moderate.

MOST LIKELY TYPE OR CONSEQUENCE OF DAMAGE FOR
INADEQUATELY PROTECTED EQUIPMENT

● Toppled lockers.
● General cleanup required.

FIGURE 3.145.   An example of personnel lockers that have been properly installed. They are base anchored at regular intervals to their foundation.

Miscellaneous Equipment

Storage, Ad Hoc

Ad hoc storage (Figure 3.146) generally does not fare well during an earthquake.

EQUIPMENT SEISMIC CATEGORY

- "E" miscellaneous equipment.

SEISMIC SPECIFICATION

- SDS-2.

SEISMIC QUALIFICATION APPROACH

- Design team judgment.

REFERENCE FIGURES FOR INSTALLATION DETAILS

- 4.102, 4.103, 4.104.

RELATIVE DEGREE OF DAMAGE OF INADEQUATELY PROTECTED EQUIPMENT

- Minor to moderate.

MOST LIKELY TYPE OR CONSEQUENCE OF DAMAGE FOR
INADEQUATELY PROTECTED EQUIPMENT

- Toppled storage.
- Spilled shelf contents.
- General cleanup required.

REFERENCE FIGURE FOR EXAMPLE OF DAMAGED EQUIPMENT

- 3.197.

FIGURE 3.146.   Example of ad hoc storage that is likely to collapse during an earthquake.

Miscellaneous Equipment

Storage Shelves

General storage shelves (Figure 3.147) must be considered with respect to the shelving as well as the shelf contents.

EQUIPMENT SEISMIC CATEGORY

- "E" miscellaneous equipment.

SEISMIC SPECIFICATION

- SDS-2.

SEISMIC QUALIFICATION APPROACH

- Equivalent static coefficient analysis.
  - Base anchorage.
  - Top bracing and anchorage.
- Design team judgment.
  - Provide shelved item restrainers.

REFERENCE FIGURES FOR INSTALLATION DETAILS

- 4.53, 4.54, 4.55, 4.56, 4.102, 4.103, 4.104.

RELATIVE DEGREE OF DAMAGE OF INADEQUATELY PROTECTED EQUIPMENT

- Minor to moderate.

MOST LIKELY TYPE OR CONSEQUENCE OF DAMAGE FOR
INADEQUATELY PROTECTED EQUIPMENT

- Shelf units topple.
- Items fall from shelves.
- General cleanup required.

REFERENCE FIGURES FOR EXAMPLES OF DAMAGED EQUIPMENT

- 3.191, 3.192, 3.193, 3.198.

FIGURE 3.147. General storage shelves such as those shown here require base anchorage, longitudinal bracing, top bracing, and shelf parapets. This shelving unit has not received any of this protection.

Miscellaneous Equipment

Typewriters

Although typewriters are not necessary for the operation of any facility, they do pose a threat to facility personnel if they fall and can add to the general cleanup required if they are not considered.

EQUIPMENT SEISMIC CATEGORY

- "E" miscellaneous equipment.

SEISMIC SPECIFICATION

- SDS-2.

SEISMIC QUALIFICATION APPROACH

- Design team judgment.
  - Anchor to desk.
  - Provide lips on desk to prevent their sliding off.

REFERENCE FIGURE FOR INSTALLATION DETAILS

- 4.110.

RELATIVE DEGREE OF DAMAGE OF INADEQUATELY PROTECTED EQUIPMENT

- Minor.

MOST LIKELY TYPE OR CONSEQUENCE OF DAMAGE FOR
INADEQUATELY PROTECTED EQUIPMENT

- General cleanup required.
- Possibility for personnel injury.

EXAMPLES OF DAMAGED EQUIPMENT

This section is included to illustrate the potential for damage to equipment by an earthquake. The preceding section of this chapter contains references to this section under the individual equipment items wherever possible.

All the example damage photographs are of equipment failures that resulted from one of three Southern California earthquakes: the San Fernando earthquake (1971, M 6.6), the Santa Barbara earthquake (1978, M 5.1), and the Imperial County earthquake (1979, M 6.6). The reader will note that these are relatively moderate earthquakes and yet the damage in many cases is quite striking. A larger magnitude earthquake produces the same types of damage, only over a larger area.

A thorough study of these photographs will give the designer, manufacturer, facility owner, and so on a much clearer picture of what to expect if building equipment is not adequately protected. These photographs address the structural integrity of equipment rather than its operational capabilities. Equipment that must remain operational may look good after an earthquake, but because of internal failures or support equipment failures may be completely inoperable. This concept must be borne in mind, especially when dealing with critical equipment.

It is hoped that the review of these photographs will underscore the importance of seismic qualification programs for all types of equipment, not just selected items that are specifically required by building codes. Designers and manufacturers have the opportunity to use this section on new facilities, while facility owners can use it as a checklist on walk-through tours of their existing buildings to improve the chance for survival of their equipment. Many existing facilities desperately need such backfitting programs even if they are not required by the existing codes.

The author wishes to apologize for the quality of some of the photographs contained in this section. They have been taken from various sources and in some cases the originals could not be located, which necessitated copying them directly from the original publication, a less than desirable procedure. It was felt, however, that the points that they illustrate are important enough to justify their reproduction.

FIGURE 3.148. Air grill that dislodged and fell. Positive attachment and a safety wire would have prevented this potentially dangerous situation. Photograph courtesy of Richard Miller and the National Science Foundation.

FIGURE 3.149. Separation of air plenum as viewed from inside the ducting. Photograph courtesy of Richard Miller and the National Science Foundation.

FIGURE 3.150. Damage to wall where inadequately restrained ducting pounded during the earthquake. This photograph illustrates the need for separating equipment that could collide during an earthquake with the potential for damage to critical equipment by support equipment. Photograph courtesy of Richard Miller and the National Science Foundation.

FIGURE 3.151. Unrestrained air handling unit slab that shifted. These slabs must be restrained if the equipment they support is to remain operational. Photograph courtesy of Richard Miller and the National Science Foundation.

FIGURE 3.152. Air bar diffuser and flexible connectors dropped from damaged ceilings. . . . Safety wires could have prevented excessive dropping of this type of equipment. Photograph courtesy of Hayakawa Associates and U.S. Department of Commerce, N.O.A.A.

FIGURE 3.153. Circular diffusers dropped from ceiling. . . . As in Figure 3.152, safety wires could have prevented this type of failure. Photograph courtesy of Hayakawa Associates and U.S. Department of Commerce, N.O.A.A.

FIGURE 3.154. Counterweight and buffer came out of the guide rails. The buffer absorbs shock at the pit bottom in case of over travel. Photograph courtesy of Leon Stein, Office of the State Architect, California.

FIGURE 3.155. Top view of the counterweight in Figure 3.154. A dislodged rail bracket can be seen at the center of the photograph behind the weights. Photograph courtesy of Leon Stein, Office of the State Architect, California.

FIGURE 3.156. Dislodged counterweight showing where it struck the bottom of an elevator car. Photograph courtesy of Leon Stein, Office of the State Architect, California.

FIGURE 3.157. Displaced counterweight rail bracket. Note the damaged beam at lower center. Photograph courtesy of Leon Stein, Office of the State Architect, California.

FIGURE 3.158. Damaged counter-weight roller guide that was removed from the elevator shaft. Photograph courtesy of Richard Miller and the National Science Foundation.

FIGURE 3.159. "Damaged controller panels and generators thrown off their mounts. . . ." Both the control panels and motor generators should have been anchored. Photograph courtesy of Hayakawa Associates and U.S. Department of Commerce, N.O.A.A.

FIGURE 3.160. Displaced motor generator. Elastomeric feet without proper anchorage will not prevent movement. Photograph courtesy Leon Stein, Office of the State Architect, California.

FIGURE 3.161. "Closeup view of generator vibration isolator disassembled by earthquake." These high-frequency vibration isolators are commonly bolted to the motor generator but not the floor. Photograph courtesy of Hayakawa Associates and U.S. Department of Commerce, N.O.A.A.

FIGURE 3.162. This unsecured hydraulic control unit shifted to the left here. Note the displacement with respect to the oil baffle and its support. Photograph courtesy of Leon Stein, Office of the State Architect, California.

FIGURE 3.163. Fire extinguishers typically fall out of their cabinets unless they are provided with quick-release catches or the doors are equipped with positive latches. Safety glass or plastic should be used for viewing panes. Photograph courtesy of Richard Miller and the National Science Foundation.

FIGURE 3.164. Spills such as that shown here can be expected wherever items are stored in cupboards without positive latches or on counter tops without some type of restraint. Photograph courtesy of Richard Miller and the National Science Foundation.

FIGURE 3.165. Fallen suspended fluorescent light fixture. Photograph courtesy of Leon Stein, Office of the State Architect, California.

FIGURE 3.166. Lay-in panel failure. Note the missing linear surface-mounted fluorescent fixtures. Photograph courtesy of Leon Stein, Office of the State Architect, California.

FIGURE 3.167. Suspended fluorescent fixtures. Note the dislodged tubes that were prevented from falling by the safety grills, which slid on the light fixture. Photograph courtesy of Richard Miller and the National Science Foundation.

FIGURE 3.168. Surface-mounted incandescent fixture that fell and was left suspended by the electrical conduit. Photograph courtesy of Richard Miller and the National Science Foundation.

FIGURE 3.169. Shelved items that were not damaged during an earthquake because of plastic shelf parapets. Photograph courtesy of Leon Stein, Office of the State Architect, California.

FIGURE 3.170. Damaged utility equipment that was left unrestrained on a portable cart. Photograph courtesy of Leon Stein, Office of the State Architect, California.

FIGURE 3.171. Typical damage that can be expected in a laboratory if the glassware and so forth is not protected. Photograph courtesy of Richard Miller and the National Science Foundation.

FIGURE 3.172.   Wall-mounted televi-
sion set that fell. Photograph courtesy of
Richard Miller and the National Science
Foundation.

FIGURE 3.173.   Collection of vibration
isolators that failed as a result of not
being provided with motion restraints.
The vibration isolator housing should not
be constructed of brittle materials. Photograph courtesy of Richard Miller and
the National Science Foundation.

FIGURE 3.174. Fume hood fans that were supplied with vibration isolators but not motion restraints. Photograph courtesy of Richard Miller and the National Science Foundation.

FIGURE 3.175. Detail of the failure from Figure 3.174. Photograph courtesy of Richard Miller and the National Science Foundation.

FIGURE 3.176. Small vibration isolator just at the point where the spring is ready to be dislodged. Photograph courtesy of Pat Lama, Mason Industries, Inc.

FIGURE 3.177. Failure of large vibration isolators at the cap screw. Photograph courtesy of J. R. Harris, National Bureau of Standards.

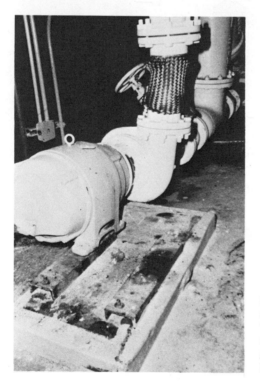

FIGURE 3.178. Pump motor that shifted. Note the braided pipe connector. This type of flexible coupling is not intended to resist axial loading. Photograph courtesy of Richard Miller and the National Science Foundation.

FIGURE 3.179. Pipe support stanchions should be anchored and not constructed of brittle material as is this broken stanchion shown here. Photograph courtesy of Richard Miller and the National Science Foundation.

FIGURE 3.180. Broken concrete pipe support stanchion. Anchors need considerable concrete coverage to prevent their failure. Photograph courtesy of Richard Miller and the National Science Foundation.

FIGURE 3.181. Pipe support with rollers to allow for expansion. Lateral bracing of the stanchion could have prevented this failure. Photograph courtesy of Pat Lama, Mason Industries, Inc.

FIGURE 3.182. Vertical pipe hanger (without provisions for lateral bracing) failure. Photograph courtesy of Pat Lama, Mason Industries, Inc.

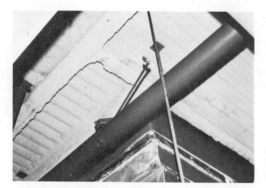

FIGURE 3.183. Broken C-clamp vertical pipe hanger without lateral support. Photograph courtesy of Leon Stein, Office of the State Architect, California.

FIGURE 3.184. These 8 inch × 48 inch × 18 gauge ± metal strips dislodged and fell to the seats below. Photograph courtesy of Leon Stein, Office of the State Architect, California.

FIGURE 3.185. Perimeter failure of suspended ceilings is common where perimeter support is not provided. Photograph courtesy of Leon Stein, Office of the State Architect, California.

FIGURE 3.186. Damaged water softener. Photograph courtesy of Hayakawa Associates and U.S. Department of Commerce, N.O.A.A.

FIGURE 3.187. Damaged water softeners. Photograph courtesy of Hayakawa Associates and U.S. Department of commerce, N.O.A.A.

FIGURE 3.188. Bolts were sheared and nuts pulled loose on base plate supporting water softener. Pipe legs supporting tanks seldom perform well. Photograph courtesy of Hayakawa Associates and U.S. Department of Commerce, N.O.A.A.

FIGURE 3.189. Damaged domestic hot water tank. Photograph courtesy of Hayakawa Associates and U.S. Department of Commerce, N.O.A.A.

FIGURE 3.190. High-pressure gas cylinders that were supported by lightweight chains. The chains were not adequately anchored to the wall in the background. Photograph courtesy of Richard Miller and the National Science Foundation.

FIGURE 3.191. Spilled files such as these can be prevented by shelf retainers. Photograph courtesy of J. R. Harris, National Bureau of Standards.

FIGURE 3.192. Boxes that spilled from their shelves. Photograph courtesy of Richard Miller and the National Science Foundation.

FIGURE 3.193. Typical library scene following an earthquake where shelf protection is not provided. Photograph courtesy of Richard Miller and the National Science Foundation.

FIGURE 3.194. Without adequate protection filing cabinets can easily topple with the potential for blocking exit paths, and so forth. Photograph courtesy of J. R. Harris, National Bureau of Standards.

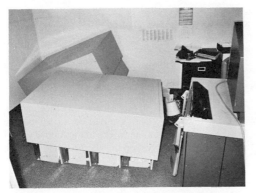

FIGURE 3.195. Top-heavy filing cabinets in the work space pose a significant hazard to personnel. Photograph courtesy of J. R. Harris, National Bureau of Standards.

FIGURE 3.196. Even items such as library card files pose a potential threat to personnel. Photograph courtesy of Richard Miller and the National Science Foundation.

FIGURE 3.197. Typical office scene where items are stored in a loosely stacked manner. Photograph courtesy of Richard Miller and the National Science Foundation.

FIGURE 3.198. Damage to ad hoc storage. Photograph courtesy of Richard Miller and the National Science Foundation.

FIGURE 3.199. Failure of dynamically sensitive internal components of a control cabinet during biaxial seismic testing. Photograph courtesy of Wyle Laboratories Scientific Services and Systems Group.

FIGURE 3.200.   Failure of hydraulic control module during biaxial seismic testing. Photograph courtesy of Wyle Laboratories Scientific Services and Systems Group.

FIGURE 3.201.   Failure of equipment rack during biaxial seismic testing. Photograph courtesy of Wyle Laboratories Scientific Services and Systems Group.

4

Anchorage and Installation Details

This chapter includes diagrams of example design installation methods for various types of equipment. As in Chapter 3, the subject is approached from the systems point of view. Each real installation may have its own characteristics and the suggested installation details may need modification for adequate assurance of survival. Where the installation details were developed by other authors, due credit is given to the earliest known author wherever possible. Other installation techniques are commonly found in construction practice and the author does not feel that any credit should be given. They are presented for the benefit of architects, engineers, and facility operators for use in new construction, as well as for those who may wish to upgrade existing facilities or those who are not familiar with all the commonly used details. The anchorage and bracing conditions are shown diagrammatically only. Each piece of equipment requires the entire seismic qualification process to adequately detail the installation. The use of these details will not guarantee equipment survivability and the details are not intended to represent such.

The reader will on some occasions notice that referrals from Chapter 3 lead to equipment installation details for other equipment. This approach is used to save space where the installation details are similar for different pieces of equipment. An example of such a case would be details for operating room lights and overhead X-ray equipment. Chapter 3 has entries for both pieces of equipment that refer the reader only to operating room lights in Chapter 4 because the installations are similar.

AIR HANDLING SYSTEMS

Guidelines for the installation of ducting have been prepared for the Sheet Metal Fund of Los Angeles (Sheet Metal and Air Conditioning Contractors' National Association, Inc., SMACNA) by the structural engineering firm of Hillman, Biddison & Loevenguth. They are titled *Guidelines for Seismic*

Restraints of Mechanical Systems and have been approved by the California Office of the State Architect. These guidelines are reproduced in part in Appendix 3, to which the reader is referred for appropriate installation details.

DATA PROCESSING SYSTEMS

Anchorage installation details for data processing systems can be extrapolated from similar applications, such as those shown in the following sections:

- Access Floor Systems.
- Communication Systems.
- Emergency Power Supply Systems.

KITCHEN SYSTEMS

Guidelines for the installation of kitchen equipment have been prepared for the Sheet Metal Industry Fund of Los Angeles (Sheet Metal and Air Conditioning Contractors' National Association, Inc., SMACNA) by the structural engineering firm of Hillman, Biddison & Loevenguth. They are titled *Guidelines for Seismic Restraints of Kitchen Equipment* and have been approved by the California Office of the State Architect. The guidelines are reproduced in part in Appendix 3, to which the reader is referred for installation details of kitchen equipment.

MOTION RESTRAINT SYSTEMS

The design team should, on every occasion, refer to the professional engineering staff of the motion restraint manufacturers when these items are required. Inappropriate installation of the motion restraints can lead to a false sense of security. Manufacturers of motion restraints are listed in Appendix 2 for the design team's easy reference.

Installing equipment that requires vibration isolation without motion restraints leads to equipment failure and can cause personnel injury or damage to other critical equipment when the springs "fly" out of their supports. Chapter 3 shows numerous occasions where cap screws sheared and spring housings broke, freeing the springs from beneath the equipment with a tremendous release of energy.

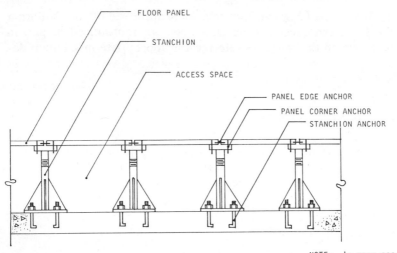

FIGURE 4.1.   Access floor systems. Interlocking access floors showing floor panels, anchors, and stanchions.

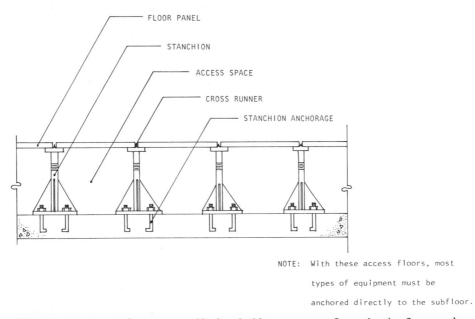

FIGURE 4.2.   Access floor systems. Noninterlocking type access floors showing floor panels and stanchions.

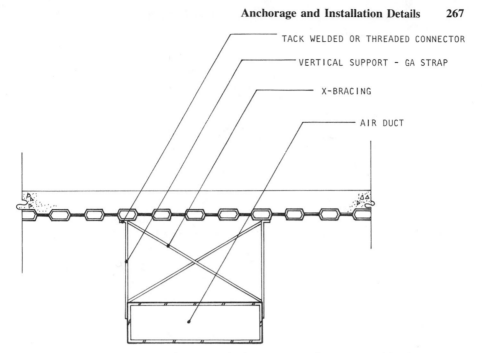

FIGURE 4.3.   Air handling systems. Air duct suspension and lateral bracing.

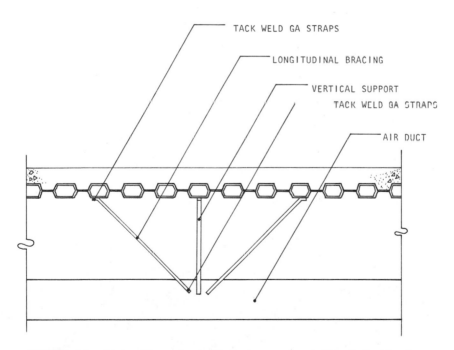

FIGURE 4.4.   Air handling systems. Air duct suspension and longitudinal bracing.

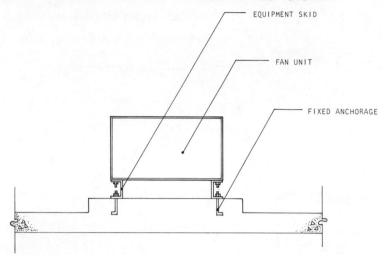

EQUIPMENT SKID

FAN UNIT

FIXED ANCHORAGE

NOTE: If vibration isolation is desired,
refer to manufacturers of vibration
isolators and motion restraints for
their recommendations.

FIGURE 4.5. Air handling systems. Fixed anchorage for fan units.

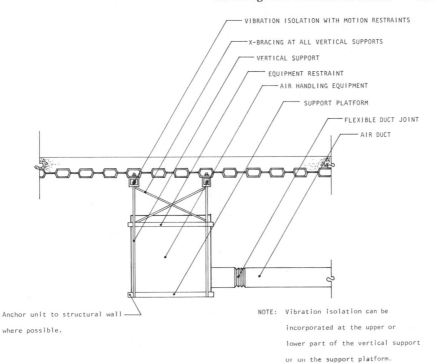

FIGURE 4.6.   Air handling systems. Suspended fan unit with vibration isolation.

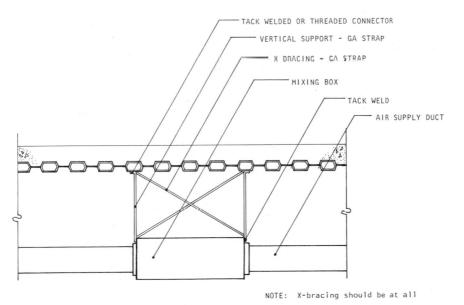

FIGURE 4.7.   Air handling systems. Suspended lightweight mixing boxes with diagonal cross (×) bracing.

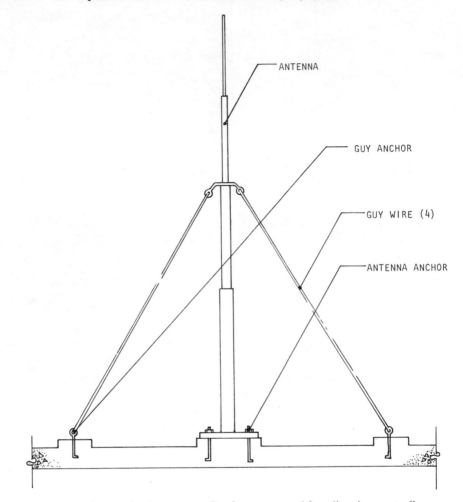

FIGURE 4.8.   Communication systems. Roof antenna guyed four directions mutually perpendicular 45° from horizontal.

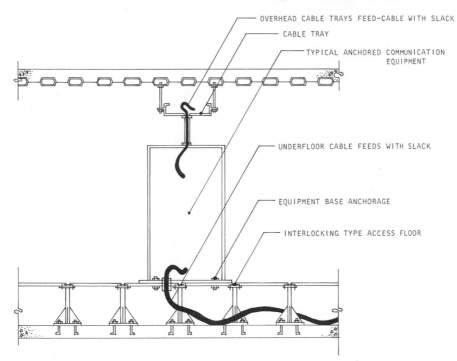

FIGURE 4.9.   Communication systems. Slack in electric cable installation.

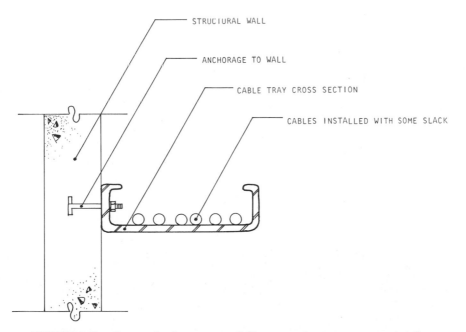

FIGURE 4.10.   Communication systems. Cable tray anchorage to structural wall.

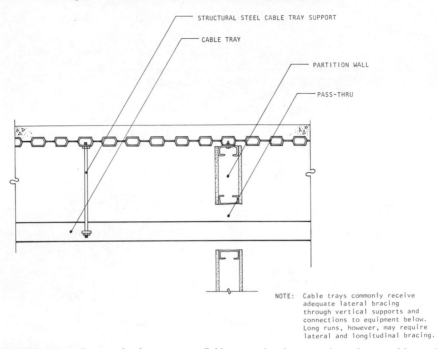

STRUCTURAL STEEL CABLE TRAY SUPPORT

CABLE TRAY

PARTITION WALL

PASS-THRU

NOTE: Cable trays commonly receive
adequate lateral bracing
through vertical supports and
connections to equipment below.
Long runs, however, may require
lateral and longitudinal bracing.

FIGURE 4.11.   Communication systems. Cable trays showing pass-through at partition wall.

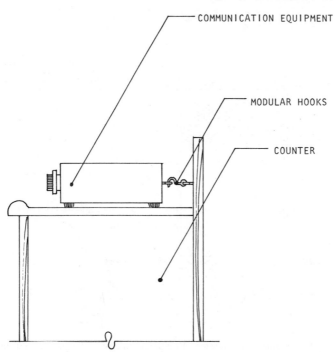

FIGURE 4.12. Communication systems. Nonpermanent counter top installation using eye hooks.

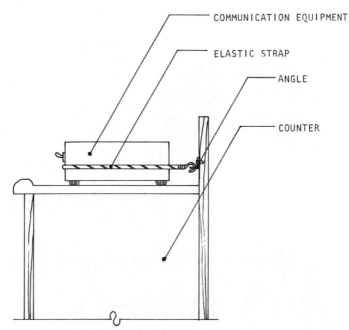

FIGURE 4.13. Communication systems. Nonpermanent counter top installation using elastic straps.

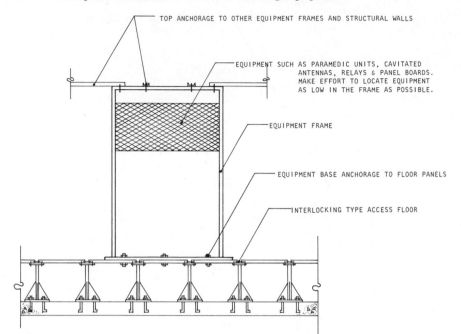

FIGURE 4.14. Communication systems. Equipment frame anchorage to interlocking type access floor.

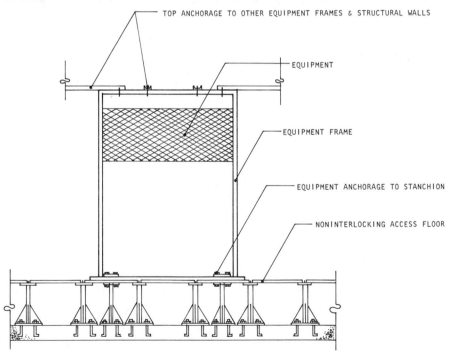

FIGURE 4.15. Communication systems. Equipment frame anchorage to noninterlocking type access floor.

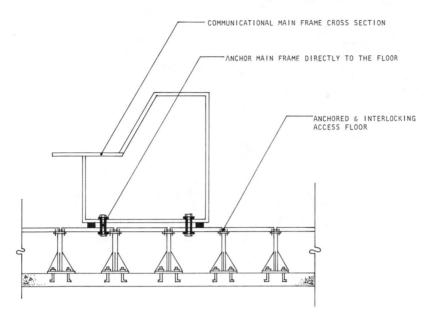

FIGURE 4.16. Communication systems. Main frame installation and anchorage for interlocking access floor.

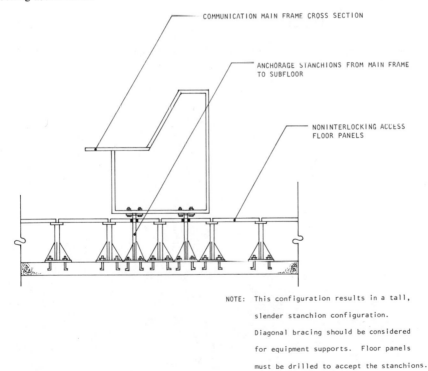

NOTE: This configuration results in a tall, slender stanchion configuration. Diagonal bracing should be considered for equipment supports. Floor panels must be drilled to accept the stanchions.

FIGURE 4.17. Communication systems. Main frame installation for noninterlocking access floor.

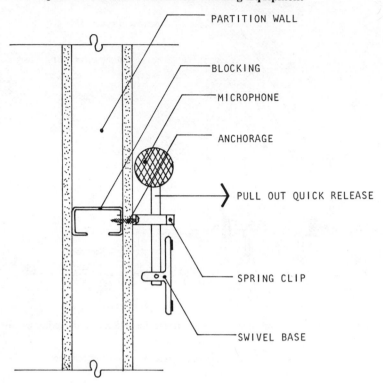

FIGURE 4.18. Communication systems. Secured microphone when not in use.

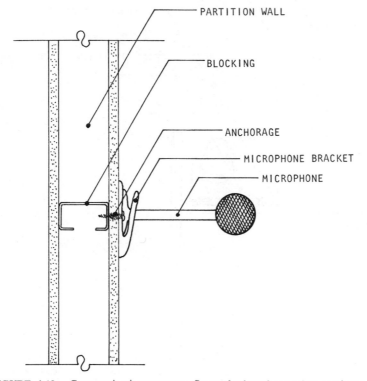

PARTITION WALL

BLOCKING

ANCHORAGE

MICROPHONE BRACKET

MICROPHONE

FIGURE 4.19. Communication systems. Secured microphone when not in use.

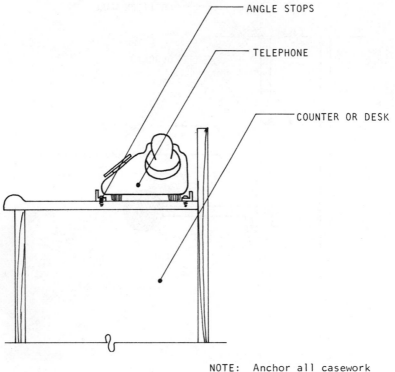

ANGLE STOPS

TELEPHONE

COUNTER OR DESK

NOTE: Anchor all casework
          to walls and/or floor

FIGURE 4.20.   Communication systems. Telephone equipment anchorage.

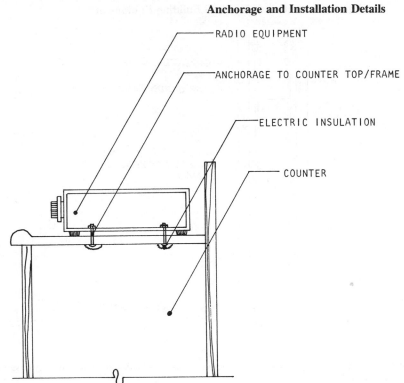

RADIO EQUIPMENT

ANCHORAGE TO COUNTER TOP/FRAME

ELECTRIC INSULATION

COUNTER

FIGURE 4.21. Communication systems. Permanent two-way radio equipment installation.

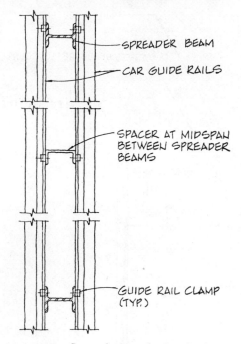

FIGURE 4.22.   Elevator systems. Spacer between back-to-back car guide rails proposed by Ayres and Sun (1973).

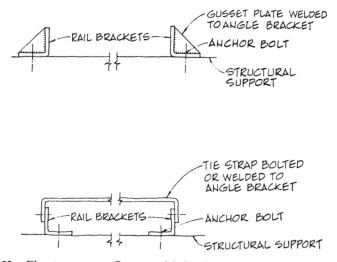

FIGURE 4.23.   Elevator systems. Counterweight brackets proposed by Ayres and Sun (1973).

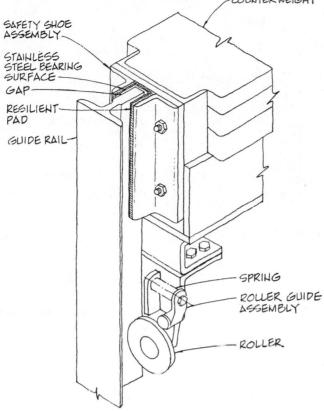

COUNTERWEIGHT

SAFETY SHOE
ASSEMBLY

STAINLESS
STEEL BEARING
SURFACE

GAP

RESILIENT
PAD

GUIDE RAIL

SPRING

ROLLER GUIDE
ASSEMBLY

ROLLER

FIGURE 4.24. Elevator systems. Counterweight safety shoe proposed by Ayres and Sun (1973).

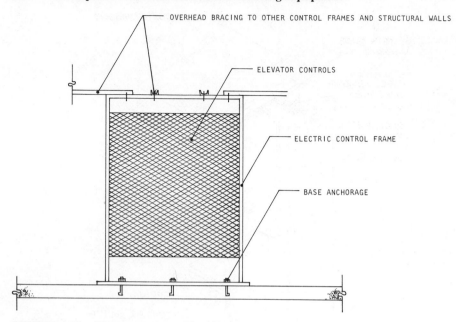

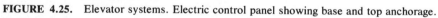

FIGURE 4.25. Elevator systems. Electric control panel showing base and top anchorage.

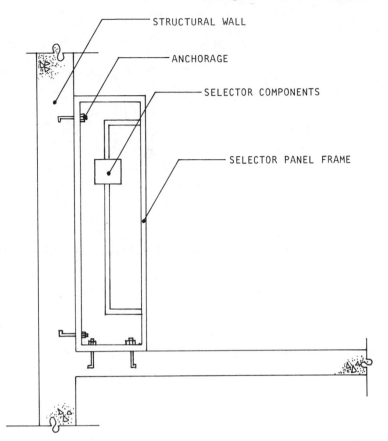

STRUCTURAL WALL

ANCHORAGE

SELECTOR COMPONENTS

SELECTOR PANEL FRAME

FIGURE 4.26.   Elevator systems. Installation of elevator selector panel.

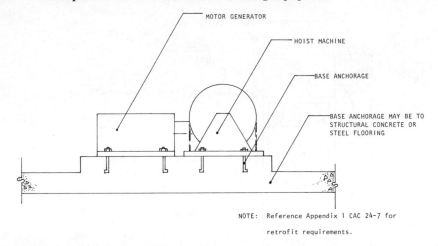

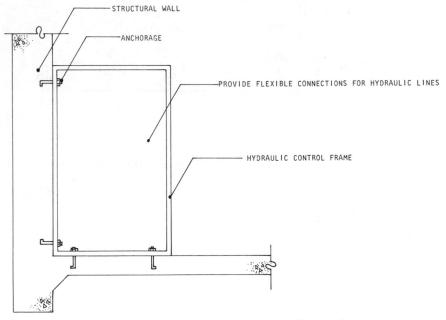

FIGURE 4.27.   Elevator systems. Installation of geared and gearless hoist machines.

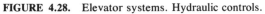

FIGURE 4.28.   Elevator systems. Hydraulic controls.

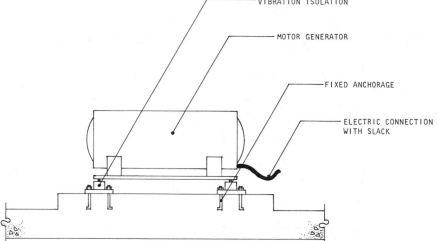

VIBRATION ISOLATION

MOTOR GENERATOR

FIXED ANCHORAGE

ELECTRIC CONNECTION
WITH SLACK

FIGURE 4.29.   Elevator systems. Anchorage of motor generators.

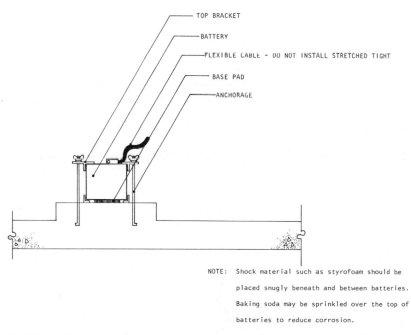

TOP BRACKET

BATTERY

FLEXIBLE CABLE – DO NOT INSTALL STRETCHED TIGHT

BASE PAD

ANCHORAGE

NOTE: Shock material such as styrofoam should be

placed snugly beneath and between batteries.

Baking soda may be sprinkled over the top of

batteries to reduce corrosion.

FIGURE 4.30.   Emergency power supply systems. Emergency power supply battery set.

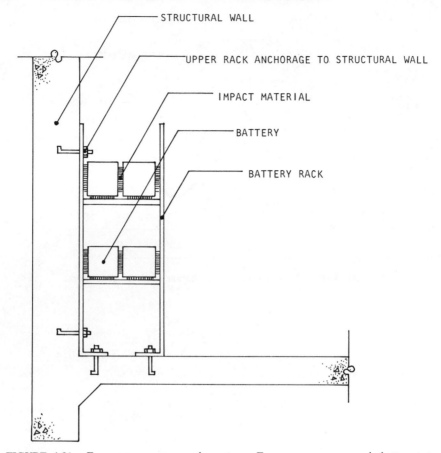

FIGURE 4.31.   Emergency power supply systems. Emergency power supply battery set.

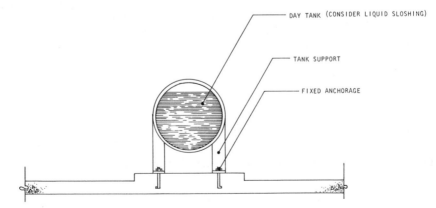

FIGURE 4.32.   Emergency power supply systems. Day tank—fixed installation.

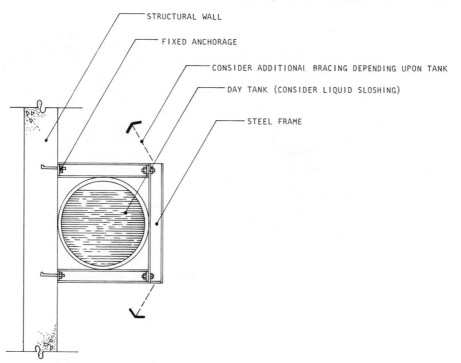

STRUCTURAL WALL

FIXED ANCHORAGE

CONSIDER ADDITIONAL BRACING DEPENDING UPON TANK

DAY TANK (CONSIDER LIQUID SLOSHING)

STEEL FRAME

FIGURE 4.33. Emergency power supply systems. Day tank—supported by a structural wall.

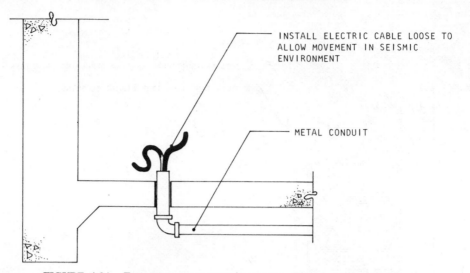

INSTALL ELECTRIC CABLE LOOSE TO
ALLOW MOVEMENT IN SEISMIC
ENVIRONMENT

METAL CONDUIT

FIGURE 4.34.   Emergency power supply systems. Electric cable installation.

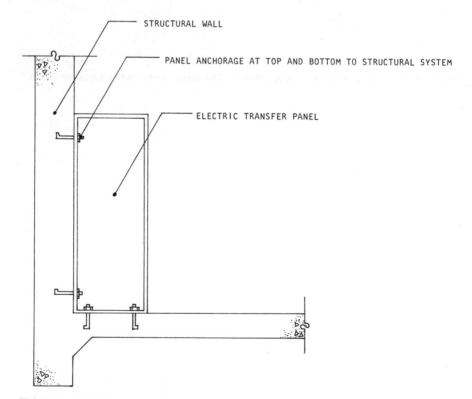

STRUCTURAL WALL

PANEL ANCHORAGE AT TOP AND BOTTOM TO STRUCTURAL SYSTEM

ELECTRIC TRANSFER PANEL

FIGURE 4.35.   Emergency power supply systems. Electric transfer panels showing frame anchorage.

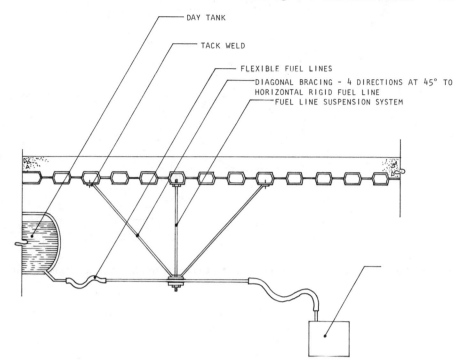

FIGURE 4.36.   Emergency power supply systems. Fuel line with flexible connections.

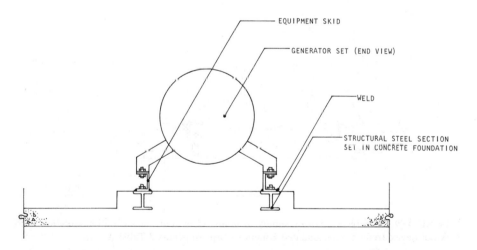

FIGURE 4.37.   Emergency power supply systems. Generator set showing fixed anchorage to structural concrete.

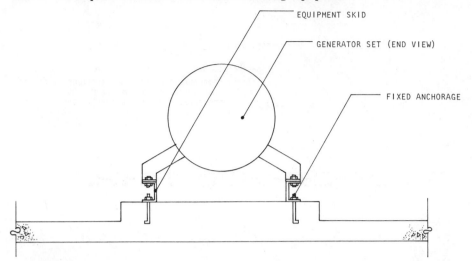

FIGURE 4.38. Emergency power supply systems. Generator set showing fixed anchorage to structural concrete.

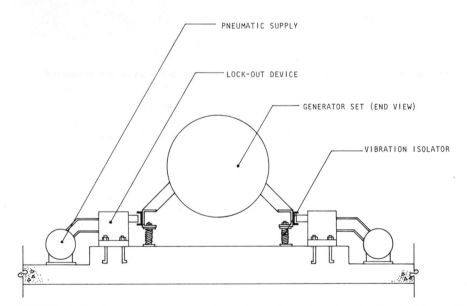

FIGURE 4.39. Emergency power supply systems. Generator set—vibration isolation with lock-out device (refer to Consolidated Kinetics Corp. Appendix 3 Table A.2.4).

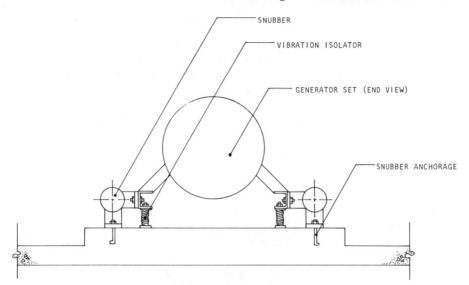

FIGURE 4.40. Emergency power supply systems. Generator set—vibration isolation with motion restraints (refer to Mason Industries Appendix 2, Table A.2.4).

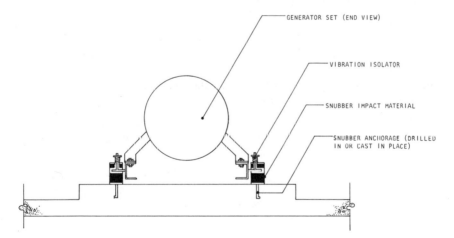

FIGURE 4.41. Emergency power supply systems. Generator set—vibration isolation with motion restraints (refer to California Dynamics Corporation Appendix 2 Table A.2.4).

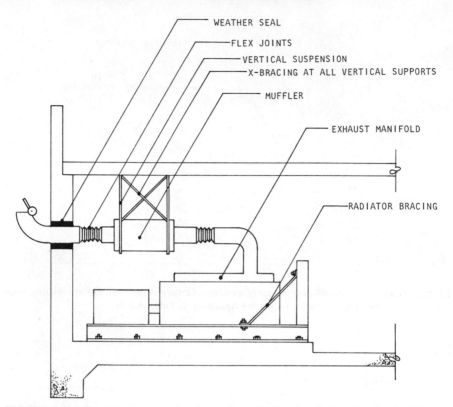

WEATHER SEAL
FLEX JOINTS
VERTICAL SUSPENSION
X-BRACING AT ALL VERTICAL SUPPORTS
MUFFLER
EXHAUST MANIFOLD
RADIATOR BRACING

FIGURE 4.42.   Emergency power supply systems. Muffler installation showing flexible con‐
nections, cross bracing, and bracing to radiator.

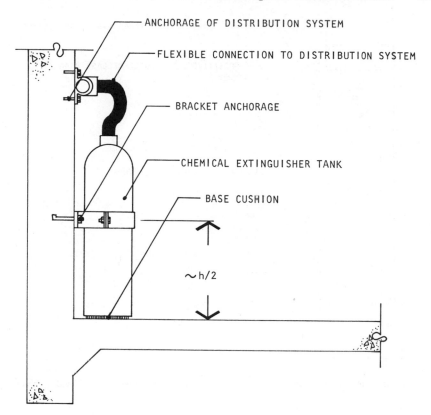

ANCHORAGE OF DISTRIBUTION SYSTEM

FLEXIBLE CONNECTION TO DISTRIBUTION SYSTEM

BRACKET ANCHORAGE

CHEMICAL EXTINGUISHER TANK

BASE CUSHION

$\sim h/2$

FIGURE 4.43. Fire protection systems. Automatic chemical extinguishing system.

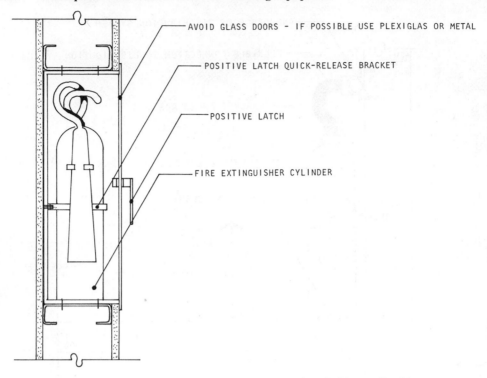

AVOID GLASS DOORS - IF POSSIBLE USE PLEXIGLAS OR METAL

POSITIVE LATCH QUICK-RELEASE BRACKET

POSITIVE LATCH

FIRE EXTINGUISHER CYLINDER

FIGURE 4.44.   Fire protection systems. Fire extinguisher installed in a wall cabinet.

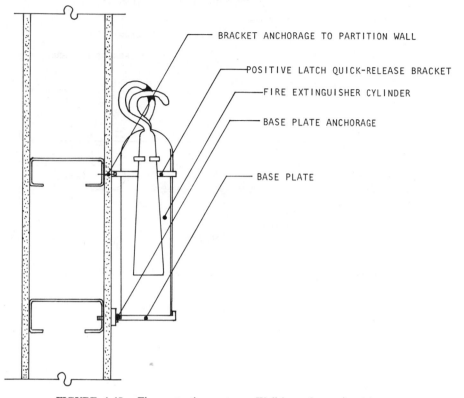

BRACKET ANCHORAGE TO PARTITION WALL

POSITIVE LATCH QUICK-RELEASE BRACKET

FIRE EXTINGUISHER CYLINDER

BASE PLATE ANCHORAGE

BASE PLATE

FIGURE 4.45.   Fire protection systems. Wall-hung fire extinguisher.

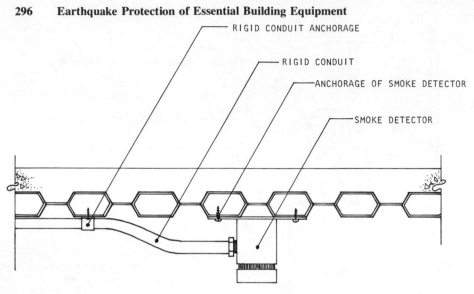

FIGURE 4.46   Fire protection systems. Smoke detector installation.

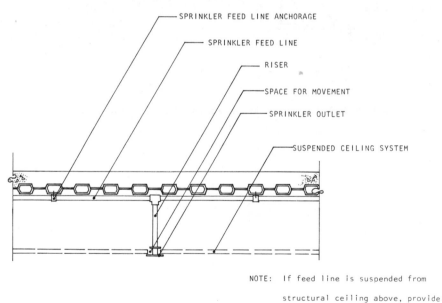

NOTE: If feed line is suspended from
         structural ceiling above, provide
         adequate lateral bracing.

FIGURE 4.47.   Fire protection systems. Sprinkler system.

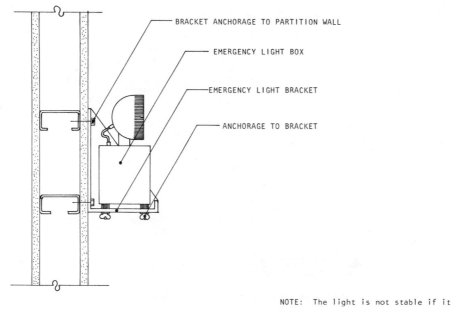

BRACKET ANCHORAGE TO PARTITION WALL

EMERGENCY LIGHT BOX

EMERGENCY LIGHT BRACKET

ANCHORAGE TO BRACKET

NOTE: The light is not stable if it

is not anchored to the brackets.

FIGURE 4.48.   Lighting Systems. Wall-mounted emergency light.

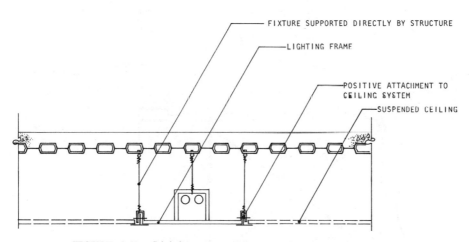

FIXTURE SUPPORTED DIRECTLY BY STRUCTURE

LIGHTING FRAME

POSITIVE ATTACHMENT TO
CEILING SYSTEM

SUSPENDED CEILING

FIGURE 4.49.   Lighting systems. Suspended ceiling lighting fixture.

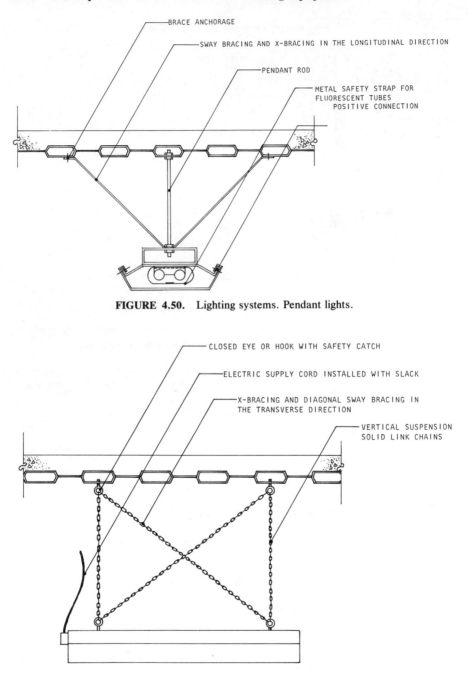

FIGURE 4.50. Lighting systems. Pendant lights.

FIGURE 4.51. Lighting systems. Suspended fluorescent lighting.

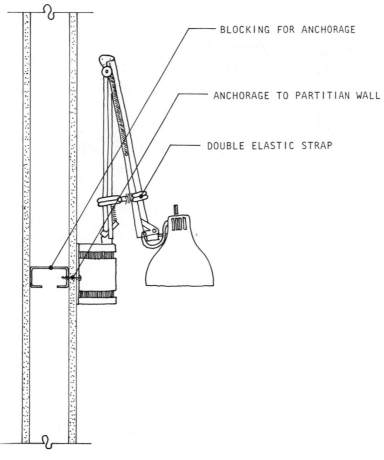

BLOCKING FOR ANCHORAGE

ANCHORAGE TO PARTITIAN WALL

DOUBLE ELASTIC STRAP

TASK LAMP

FIGURE 4.52.   Lighting systems Task lamp. Adapted from Stone, Marraccini, and Patterson, 1976.

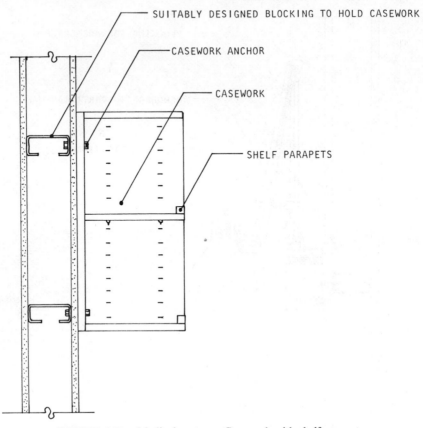

SUITABLY DESIGNED BLOCKING TO HOLD CASEWORK

CASEWORK ANCHOR

CASEWORK

SHELF PARAPETS

FIGURE 4.53. Medical systems. Casework with shelf parapets.

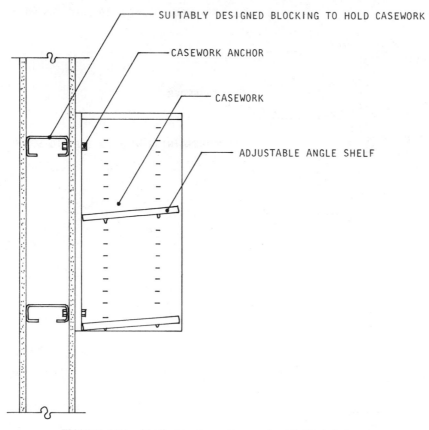

SUITABLY DESIGNED BLOCKING TO HOLD CASEWORK

CASEWORK ANCHOR

CASEWORK

ADJUSTABLE ANGLE SHELF

FIGURE 4.54.   Medical systems. Casework with tilted shelves.

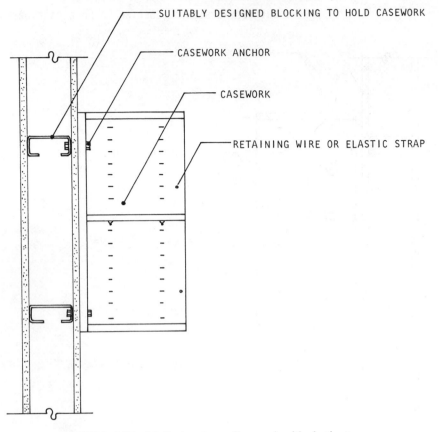

FIGURE 4.55.   Medical systems. Casework with elastic straps.

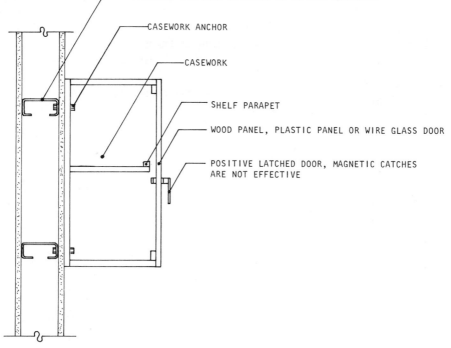

SUITABLY DESIGNED BLOCKING TO SUPPORT CASEWORK

CASEWORK ANCHOR

CASEWORK

SHELF PARAPET

WOOD PANEL, PLASTIC PANEL OR WIRE GLASS DOOR

POSITIVE LATCHED DOOR, MAGNETIC CATCHES ARE NOT EFFECTIVE

FIGURE 4.56.   Medical systems. Casework with positive latched door.

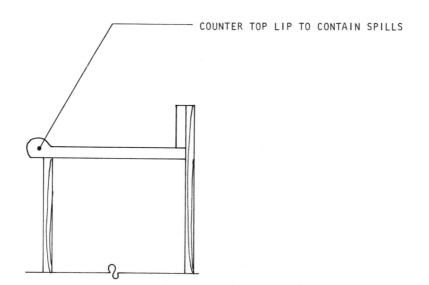

COUNTER TOP LIP TO CONTAIN SPILLS

FIGURE 4.57.   Medical systems. Counter top with lip.

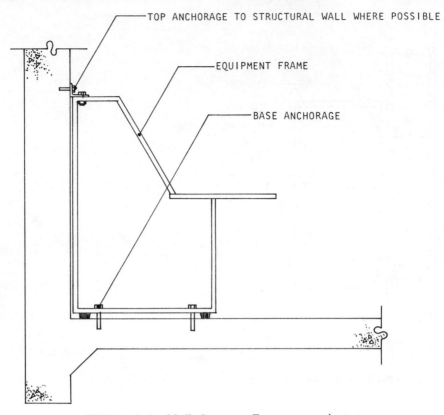

TOP ANCHORAGE TO STRUCTURAL WALL WHERE POSSIBLE

EQUIPMENT FRAME

BASE ANCHORAGE

FIGURE 4.58.   Medical systems. Frame type equipment.

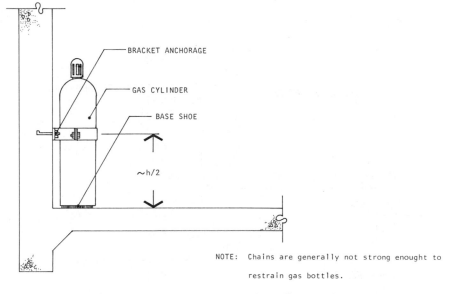

NOTE: Chains are generally not strong enought to

restrain gas bottles.

FIGURE 4.59. Medical systems. Gas cylinder anchorage.

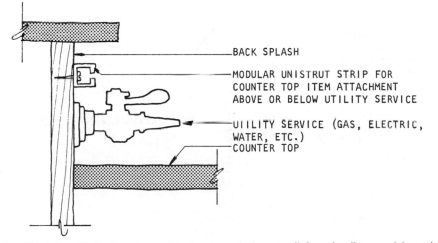

FIGURE 4.60. Medical systems. Counter top attachment rail for miscellaneous lab equipment. Adapted from Stone, Marraccini, and Patterson, 1976.

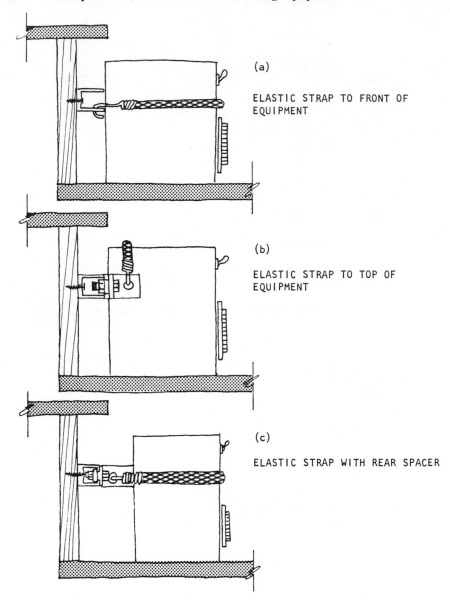

FIGURE 4.61. Medical systems. Counter top item attachment with elastic straps. Adapted from Stone, Marraccini, and Patterson, 1976.

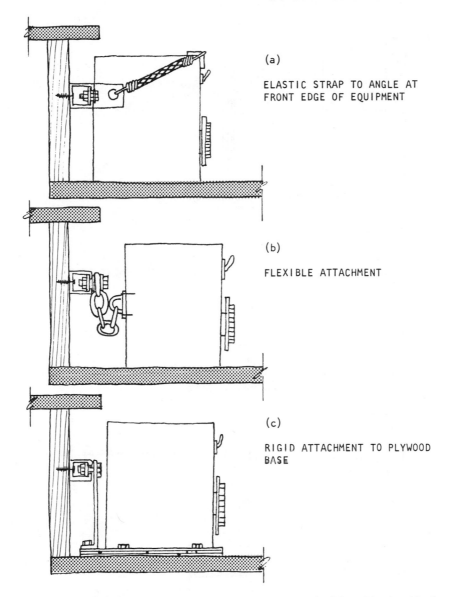

(a)

ELASTIC STRAP TO ANGLE AT
FRONT EDGE OF EQUIPMENT

(b)

FLEXIBLE ATTACHMENT

(c)

RIGID ATTACHMENT TO PLYWOOD
BASE

FIGURE 4.62. Medical systems. Counter top item attachment—flexible and fixed positioning. Adapted from Stone, Marraccini, and Patterson, 1976.

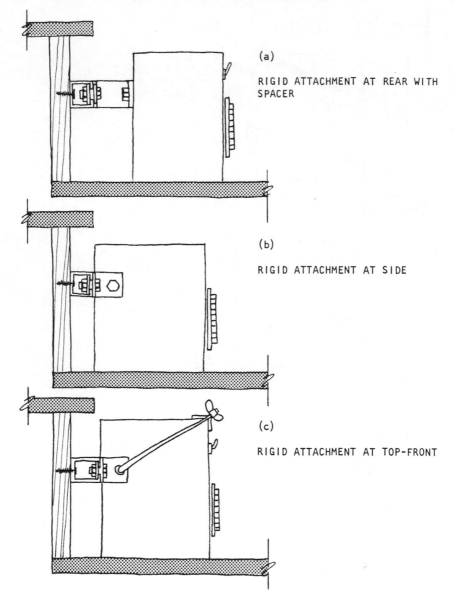

(a)

RIGID ATTACHMENT AT REAR WITH
SPACER

(b)

RIGID ATTACHMENT AT SIDE

(c)

RIGID ATTACHMENT AT TOP-FRONT

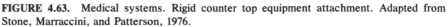

FIGURE 4.63. Medical systems. Rigid counter top equipment attachment. Adapted from Stone, Marraccini, and Patterson, 1976.

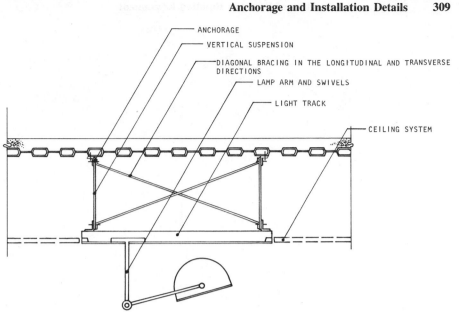

FIGURE 4.64. Medical systems. Operating room lighting.

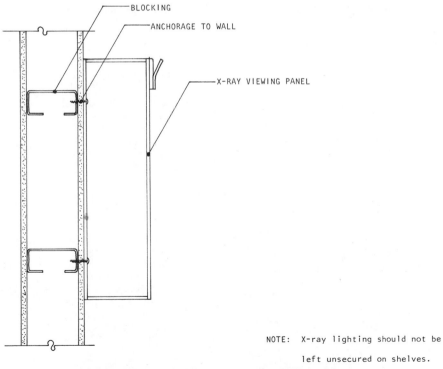

NOTE: X-ray lighting should not be

left unsecured on shelves.

FIGURE 4.65. Medical systems. X-Ray lighting anchorage.

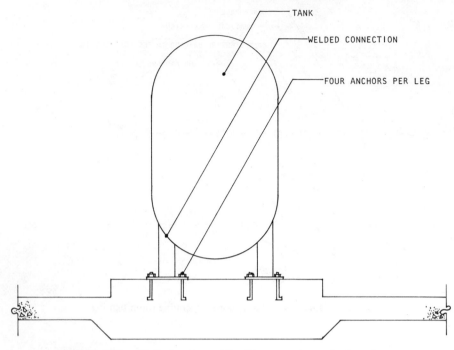

FIGURE 4.66. Medical systems. Liquid oxygen storage tank anchorage.

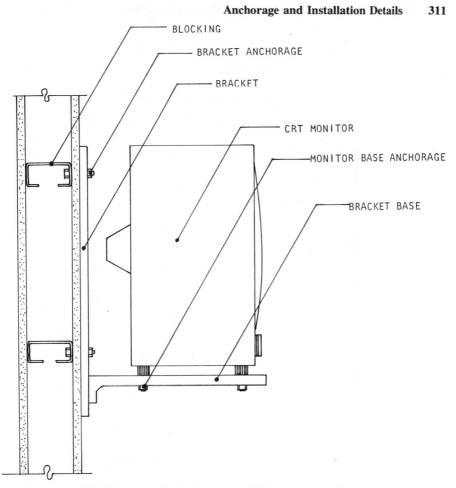

BLOCKING

BRACKET ANCHORAGE

BRACKET

CRT MONITOR

MONITOR BASE ANCHORAGE

BRACKET BASE

FIGURE 4.67.   Medical systems. CRT monitor installation.

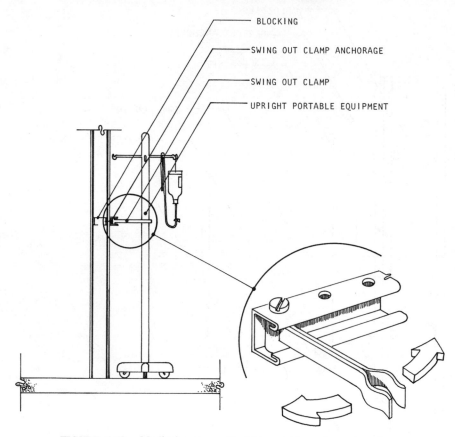

FIGURE 4.68.   Medical systems. Upright portable equipment anchorage.

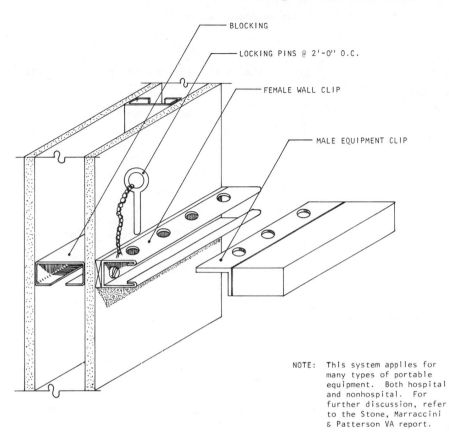

BLOCKING

LOCKING PINS @ 2'-0" O.C.

FEMALE WALL CLIP

MALE EQUIPMENT CLIP

NOTE: This system applies for many types of portable equipment. Both hospital and nonhospital. For further discussion, refer to the Stone, Marraccini & Patterson VA report.

FIGURE 4.69. Medical systems. Portable equipment retainer. Adapted from Stone, Marraccini, and Patterson, 1976.

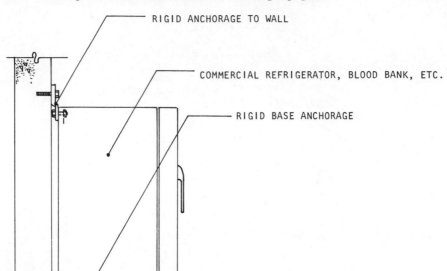

FIGURE 4.70.   Medical systems. Installation of commercial refrigerator, blood bank, bone bank, and so on.

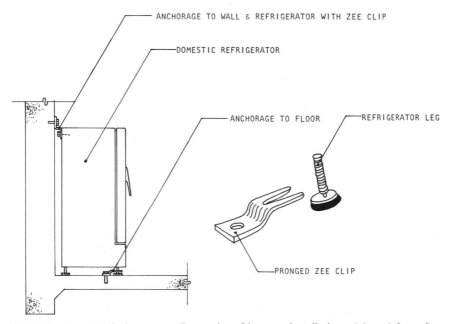

ANCHORAGE TO WALL & REFRIGERATOR WITH ZEE CLIP

DOMESTIC REFRIGERATOR

ANCHORAGE TO FLOOR

REFRIGERATOR LEG

PRONGED ZEE CLIP

FIGURE 4.71. Medical systems. Domestic refrigerator installation. Adapted from Stone, Marraccini, and Patterson, 1976.

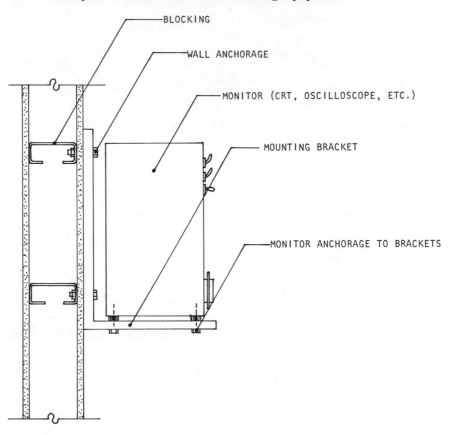

FIGURE 4.72.   Medical systems. Monitor anchorage to wall-mounted shelf.

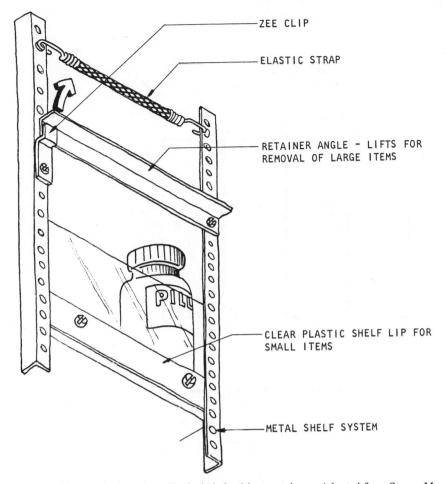

ZEE CLIP

ELASTIC STRAP

RETAINER ANGLE - LIFTS FOR
REMOVAL OF LARGE ITEMS

CLEAR PLASTIC SHELF LIP FOR
SMALL ITEMS

METAL SHELF SYSTEM

FIGURE 4.73.   Medical systems. Typical shelved item retainers. Adapted from Stone, Marraccini and Patterson, 1976.

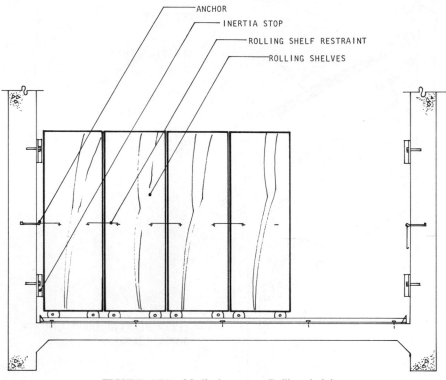

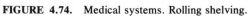

FIGURE 4.74.   Medical systems. Rolling shelving.

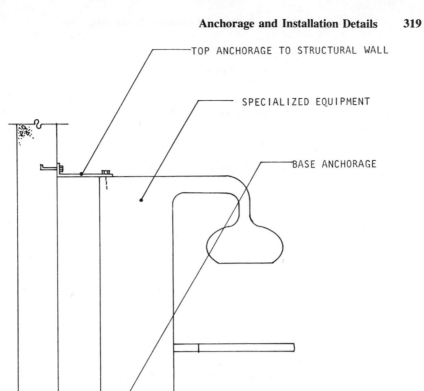

TOP ANCHORAGE TO STRUCTURAL WALL

SPECIALIZED EQUIPMENT

BASE ANCHORAGE

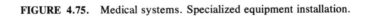

FIGURE 4.75.   Medical systems. Specialized equipment installation.

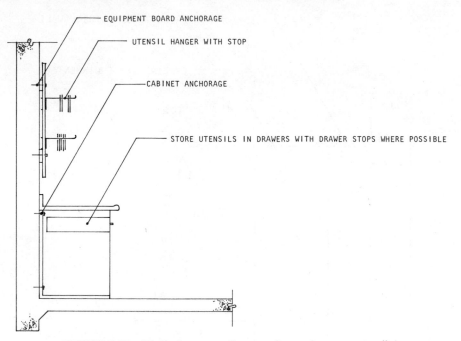

EQUIPMENT BOARD ANCHORAGE

UTENSIL HANGER WITH STOP

CABINET ANCHORAGE

STORE UTENSILS IN DRAWERS WITH DRAWER STOPS WHERE POSSIBLE

FIGURE 4.76.   Medical systems. Storage of operating room utensils.

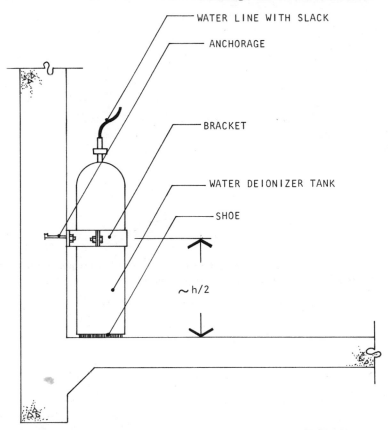

WATER LINE WITH SLACK

ANCHORAGE

BRACKET

WATER DEIONIZER TANK

SHOE

~ h/2

FIGURE 4.77. Medical systems. Storage of water deionizers for kidney dialysis.

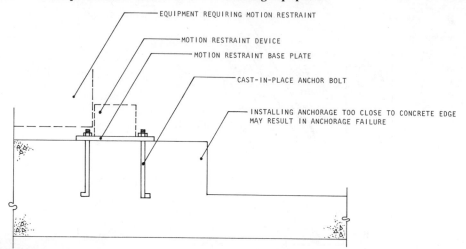

FIGURE 4.78.   Motion restraint systems. Cast-in-place anchor bolts require greater accuracy of all trades associated with installation of this equipment.

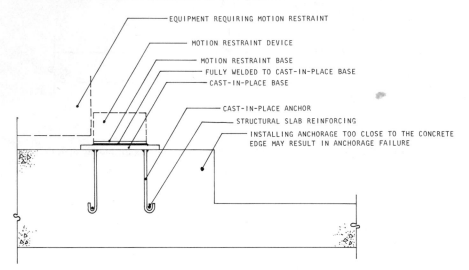

FIGURE 4.79.   Motion restraint systems. Cast-in-place base plate anchorage allows the equipment installer greater flexibility.

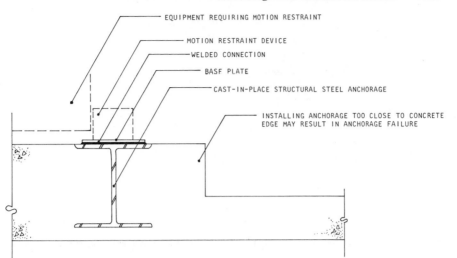

FIGURE 4.80. Motion restraint systems. Cast-in-place structural steel anchorage for motion restraint devices.

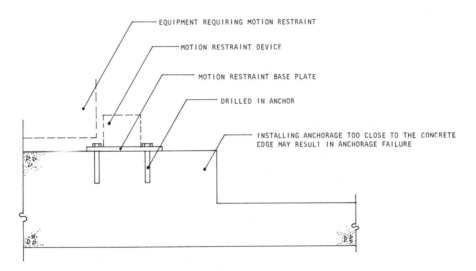

FIGURE 4.81. Motion restraint systems. Drilled in anchors allow the field trades ample latitude in equipment installation.

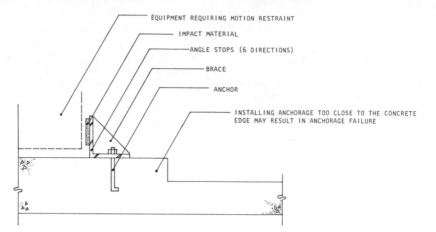

FIGURE 4.82. Motion restraint systems. Angle stops for lightweight, noncritical equipment.

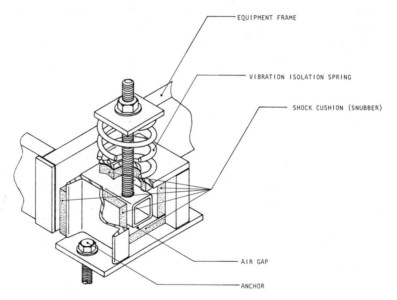

FIGURE 4.83. Motion restraint systems. Integral vibration isolation and snubbing device (California Dynamics Corporation type HQ; refer to Appendix 2, California Dynamics Corporation for installation applications and requirements.)

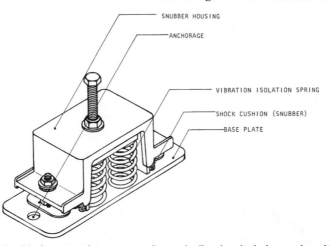

FIGURE 4.84. Motion restraint systems. Integral vibration isolation and snubbing device (California Dynamics Corporation Type RJ; refer to Appendix 2, California Dynamics Corporation for installation applications and requirements.)

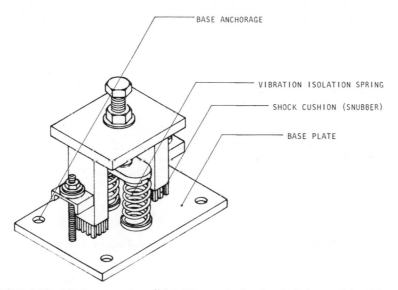

FIGURE 4.85. Motion restraint systems. Integral vibration isolation and snubbing device (California Dynamics Corporation type RJS; refer to Appendix 2, California Dynamics Corporation for installation applications and requirements.)

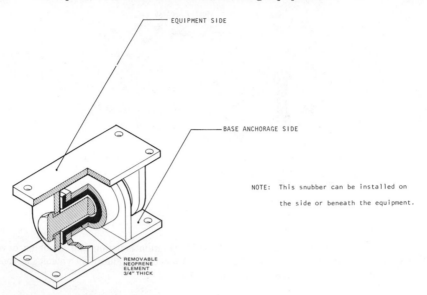

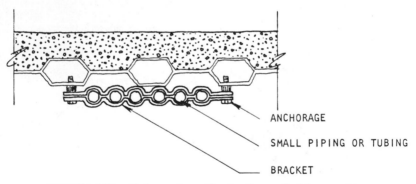

FIGURE 4.86. Motion restraint systems. Typical seismic all directional snubber. (Mason Industries type Z-1011; refer to Appendix 2, Mason Industries, Inc. for installation applications and requirements.)

FIGURE 4.87. Piping systems. Bracing for small piping or tubing.

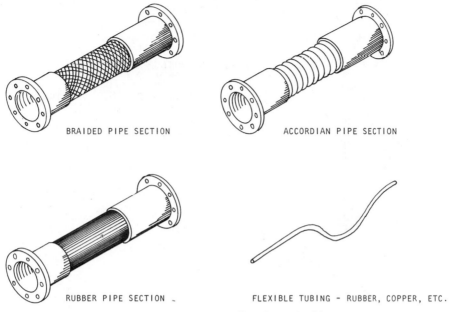

BRAIDED PIPE SECTION   ACCORDIAN PIPE SECTION

RUBBER PIPE SECTION   FLEXIBLE TUBING - RUBBER, COPPER, ETC.

FIGURE 4.88.   Piping systems. Flexible pipe and tubing connectors.

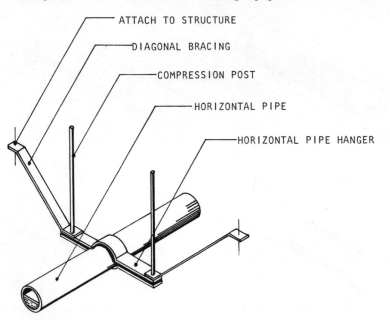

ATTACH TO STRUCTURE

DIAGONAL BRACING

COMPRESSION POST

HORIZONTAL PIPE

HORIZONTAL PIPE HANGER

FIGURE 4.89.   Piping systems. Horizontal laterally braced pipe hanger.

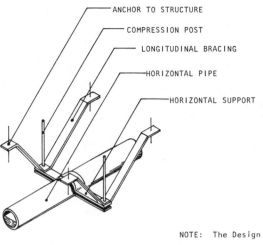

ANCHOR TO STRUCTURE

COMPRESSION POST

LONGITUDINAL BRACING

HORIZONTAL PIPE

HORIZONTAL SUPPORT

NOTE: The Design Team may wish to consider allowing
          slight longitudinal movements with damped
          restraints.

FIGURE 4.90.   Piping systems. Horizontal pipe hanger with longitudinal bracing.

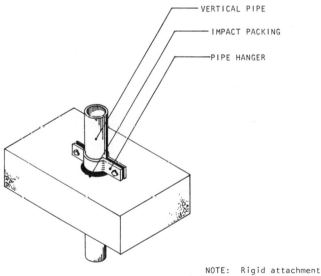

VERTICAL PIPE

IMPACT PACKING

PIPE HANGER

NOTE: Rigid attachment to the structural floor
          increases potential for pipe failure.

FIGURE 4.91.   Piping systems. Vertical pipe hanger.

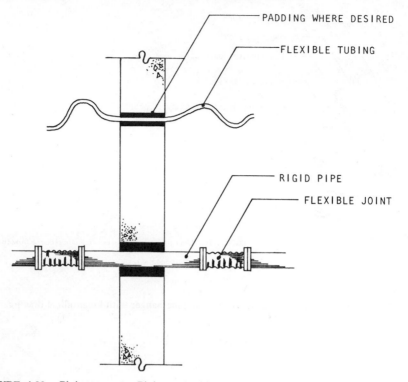

PADDING WHERE DESIRED

FLEXIBLE TUBING

RIGID PIPE

FLEXIBLE JOINT

FIGURE 4.92.   Piping systems. Piping and tubing at wall. Flexible allowances also applies to floor interfaces.

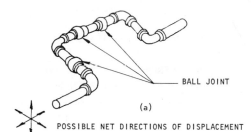

(a)

POSSIBLE NET DIRECTIONS OF DISPLACEMENT

FROM: Ayres & Sun, "Nonstructural Damage", San Fernando, CA., Earthquake of Feb. 9, 1971, Vol. 1, Part B, N.O.A.A., Washington, D.C., 1973, Page 759

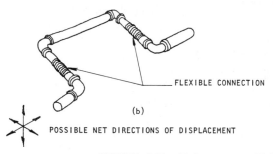

(b)

POSSIBLE NET DIRECTIONS OF DISPLACEMENT

FIGURE 4.93. Piping systems. Piping at seismic joint.

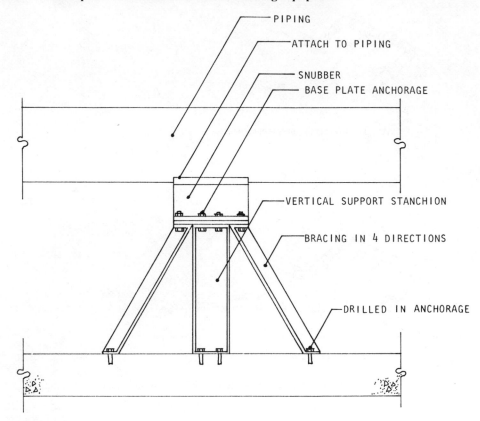

FIGURE 4.94. Piping systems. Seismic bracing for stanchion-mounted pipes. (Courtesy Mason Industries, Inc.)

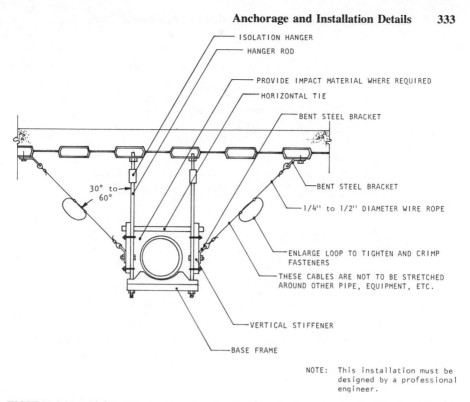

ISOLATION HANGER

HANGER ROD

PROVIDE IMPACT MATERIAL WHERE REQUIRED

HORIZONTAL TIE

BENT STEEL BRACKET

30° to 60°

BENT STEEL BRACKET

1/4" to 1/2" DIAMETER WIRE ROPE

ENLARGE LOOP TO TIGHTEN AND CRIMP FASTENERS

THESE CABLES ARE NOT TO BE STRETCHED AROUND OTHER PIPE, EQUIPMENT, ETC.

VERTICAL STIFFENER

BASE FRAME

NOTE: This installation must be designed by a professional engineer.

FIGURE 4.95.   Piping systems. Sway bracing for the seismic environment. (Courtesy Mason Industries, Inc.)

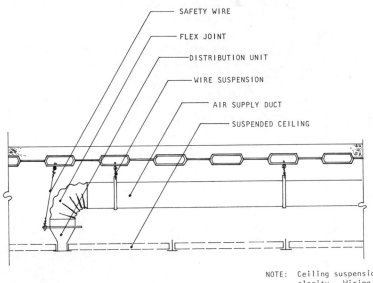

SAFETY WIRE

FLEX JOINT

DISTRIBUTION UNIT

WIRE SUSPENSION

AIR SUPPLY DUCT

SUSPENDED CEILING

NOTE: Ceiling suspension omitted for
clarity. Wiring should not wrap
or bend around ducting.

FIGURE 4.96.   Suspended ceiling systems. Air distribution for integrated ceiling system.

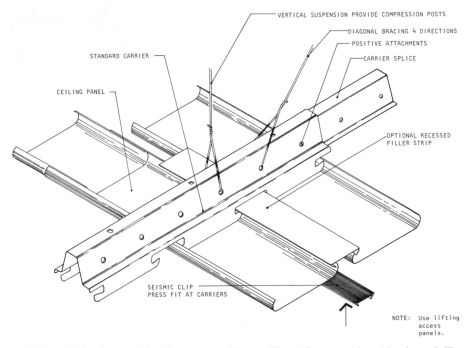

VERTICAL SUSPENSION PROVIDE COMPRESSION POSTS

DIAGONAL BRACING 4 DIRECTIONS

POSITIVE ATTACHMENTS

CARRIER SPLICE

STANDARD CARRIER

CEILING PANEL

OPTIONAL RECESSED
FILLER STRIP

SEISMIC CLIP
PRESS FIT AT CARRIERS

NOTE: Use lifting
access
panels.

FIGURE 4.97.   Suspended ceiling systems. Metal ceilings. (Courtesy Alcoa Aluminum Ceiling Systems.)

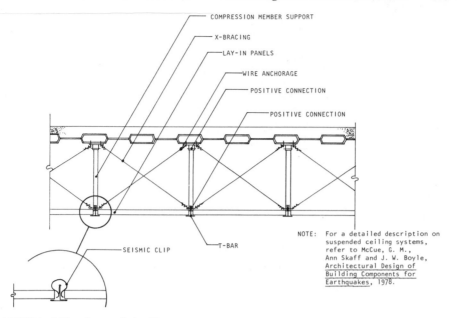

COMPRESSION MEMBER SUPPORT

X-BRACING

LAY-IN PANELS

WIRE ANCHORAGE

POSITIVE CONNECTION

POSITIVE CONNECTION

NOTE: For a detailed description on suspended ceiling systems, refer to McCue, G. M., Ann Skaff and J. W. Boyle, <u>Architectural Design of Building Components for Earthquakes</u>, 1978.

SEISMIC CLIP

T-BAR

FIGURE 4.98. Suspended ceiling systems. T-bar and acoustic lay-in ceiling suspension system.

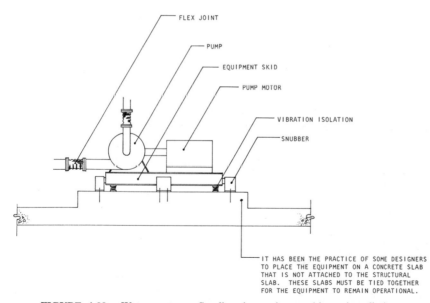

FLEX JOINT

PUMP

EQUIPMENT SKID

PUMP MOTOR

VIBRATION ISOLATION

SNUBBER

IT HAS BEEN THE PRACTICE OF SOME DESIGNERS TO PLACE THE EQUIPMENT ON A CONCRETE SLAB THAT IS NOT ATTACHED TO THE STRUCTURAL SLAB. THESE SLABS MUST BE TIED TOGETHER FOR THE EQUIPMENT TO REMAIN OPERATIONAL.

FIGURE 4.99. Water systems. Small reciprocating machinery installation.

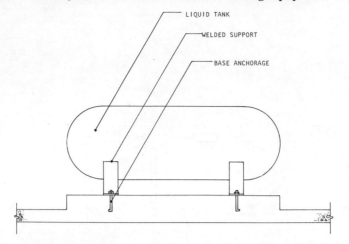

NOTE: Consider liquid sloshing in tank.

FIGURE 4.100. Water systems. Installation of horizontal liquid tanks.

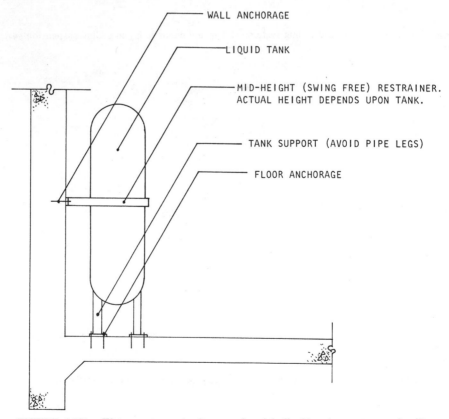

FIGURE 4.101. Water systems. Anchorage of upright liquid tanks to structural walls.

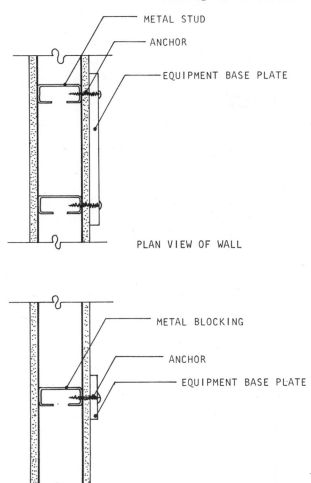

METAL STUD

ANCHOR

EQUIPMENT BASE PLATE

PLAN VIEW OF WALL

METAL BLOCKING

ANCHOR

EQUIPMENT BASE PLATE

SECTION VIEW OF WALL

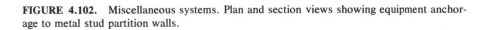

FIGURE 4.102. Miscellaneous systems. Plan and section views showing equipment anchorage to metal stud partition walls.

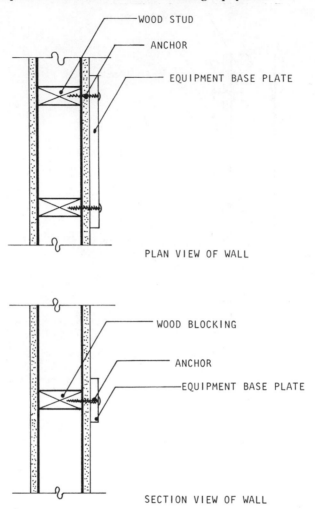

FIGURE 4.103.   Miscellaneous equipment. Plan and section views showing equipment anchorage to wood stud partition walls.

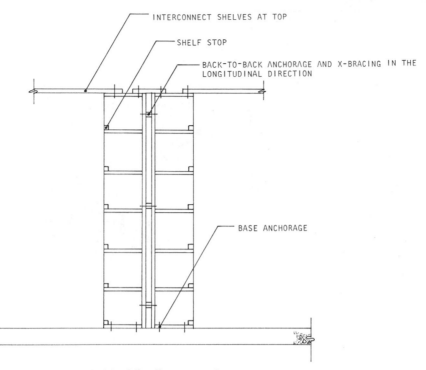

INTERCONNECT SHELVES AT TOP

SHELF STOP

BACK-TO-BACK ANCHORAGE AND X-BRACING IN THE
LONGITUDINAL DIRECTION

BASE ANCHORAGE

FIGURE 4.104.   Miscellaneous equipment. Book shelf anchorage.

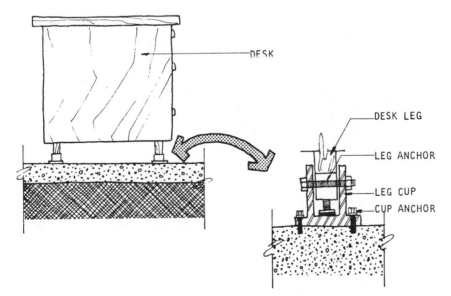

DESK

DESK LEG

LEG ANCHOR

LEG CUP

CUP ANCHOR

FIGURE 4.105.   Miscellaneous equipment. Permanent desk installation.

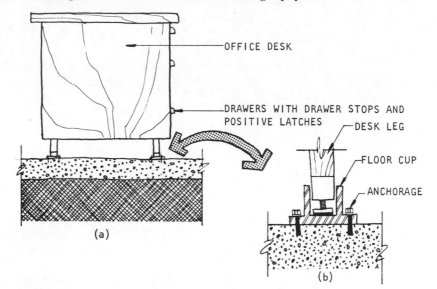

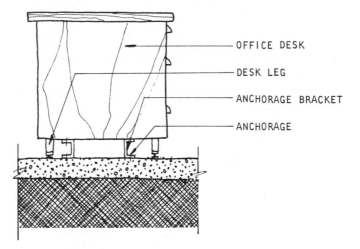

FIGURE 4.106.   Miscellaneous equipment. Nonpermanent office desk floor cups.

FIGURE 4.107.   Miscellaneous equipment. Permanent desk installation.

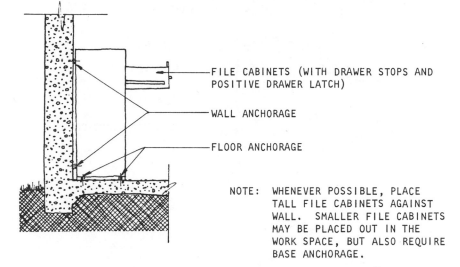

FILE CABINETS (WITH DRAWER STOPS AND POSITIVE DRAWER LATCH)

WALL ANCHORAGE

FLOOR ANCHORAGE

NOTE: WHENEVER POSSIBLE, PLACE
       TALL FILE CABINETS AGAINST
       WALL. SMALLER FILE CABINETS
       MAY BE PLACED OUT IN THE
       WORK SPACE, BUT ALSO REQUIRE
       BASE ANCHORAGE.

FIGURE 4.108.   Miscellaneous equipment. File cabinet installation.

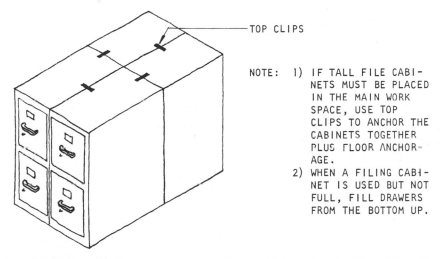

TOP CLIPS

NOTE: 1) IF TALL FILE CABI-
          NETS MUST BE PLACED
          IN THE MAIN WORK
          SPACE, USE TOP
          CLIPS TO ANCHOR THE
          CABINETS TOGETHER
          PLUS FLOOR ANCHOR-
          AGE.
       2) WHEN A FILING CABI-
          NET IS USED BUT NOT
          FULL, FILL DRAWERS
          FROM THE BOTTOM UP.

FIGURE 4.109.   Miscellaneous equipment. File cabinet interanchorage. Adapted from Stone, Marraccini and Patterson, 1976.

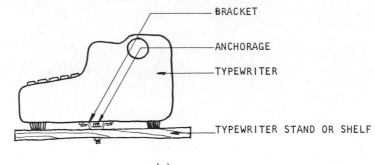

(a)

PERMANENT INSTALLATION

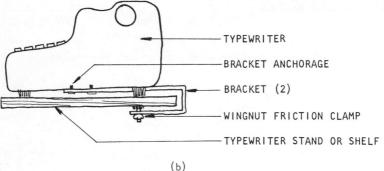

(b)

NONPERMANENT INSTALLATION

FIGURE 4.110.   Miscellaneous equipment. Typewriter installation to prevent dislocation. Adapted from Stone, Marraccini, and Patterson, 1976.

5

Concluding Remarks

There are many reasons for considering a seismic qualification program for facilities located in earthquake prone areas. These programs are required by code for essential facilities and can be equally justified by environmental and economical reasons for most other types of facilities. Building equipment is sometimes reduced to a useless pile of rubble by an earthquake even when the building structure itself is left intact. The comprehensive seismic qualification program suggested in this book is geared to reduce the potential for damage to all building equipment whether it is for a hospital, library, refinery, school, or other building. Most facilities can benefit by some measure of protection if they are located in an earthquake prone area.

The review of building codes in Appendix 1 shows that most of them typically approach the equipment qualification problem from a static point of view. While it is a valid method of qualification for many types of equipment, it is just as invalid for other types. This book presents the reader with the basics for qualification programs that utilize methods that are specifically suited to the requirements and characteristics of each individual piece of equipment. Earthquake testing is required for critical equipment that is likely to fail operationally rather than structurally. Dynamic analyses are required for equipment that is likely to wiggle extensively because of its geometry and is not likely to fail operationally. The static analysis is valid for the type of equipment that is generally rigid and only needs to remain anchored to be operational and not collide with other more critical equipment.

This book introduces the reader to a method whereby all equipment can be assigned to a seismic category, which then lends itself to examination and qualification by one of the two model seismic design specifications presented. All this information is provided for various types of equipment and is referenced to idealized diagrammatic installation details in Chapter 4 and Appendix 3. The qualification methods presented herein are intended for use by all those interested in the survivability of building equipment during and after an earthquake.

Research programs are currently needed to refine the coefficients used by codes in the static formulas. Seismic testing programs can be designed for and conducted on typical equipment installations that will yield coefficients that are more valid than those currently employed. The National Science

Foundation and some university and private testing laboratories are currently pioneering in these efforts. Upgraded codes from these findings recognize the dynamic nature, as well as the operability requirements, of equipment that will certainly result in facilities less likely to be damaged as a result of an earthquake. Examples of such endeavors include research on the nonstructural behavior of the county services building (during the 1979 Imperial Valley Earthquake) by Chris Arnold at Building Systems Development in Palo Alto and Dr. Satwant Rihal's research on the seismic performance of nonstructural partition walls at Cal-Poly San Luis Obispo. Extensive testing programs are continuously being conducted at private companies such as Wyle Laboratories seismic simulators at Norco, California and Huntsville, Alabama, and at ANCO in Santa Monica, California.

The approaches that are presented to the reader are not new and have been utilized by critical facilities such as nuclear power plants for more than a decade. The technology exists to attain a relatively high level of confidence with respect to survivability of equipment after a severe earthquake and is only waiting to be employed.

Bibliography

Anderson, Thomas L., and Douglas J. Nyman. "Lifeline Earthquake Engineering for Trans Alaska Pipeline System," *The Current State of Knowledge of Lifeline Earthquake Engineering.* American Society of Civil Engineers, U.C.L.A., 1977.

Ayres, J. Marx, T. Y. Sun, and F. R. Brown. *Report on Nonstructural Damage to Buildings Due to the March 27, 1964 Alaska Earthquake.* National Academy of Sciences, Committee on the Alaskan Earthquake Engineering Panel, 1967.

Ayres, J. Marx and Tseng-Yao Sun. "Nonstructural Damage," *San Fernando, California Earthquake of February 9, 1971.* L. M. Murphy, Ed. Volume 1, Part B. N.O.A.A., Washington, D.C., 1973.

Ayres, J. Marx, and Tseng-Yao Sun. "Criteria for Building Services and Furnishings," *Building Practices for Disaster Mitigation.* Richard Wright, Samuel Kramer, and Charles Culver, Eds., National Bureau of Standards, Science Series 46, U. S. Department of Commerce, Workshop, 1972, issued February 1973.

Basis for Seismic Restraint Design of Mechanical and Electrical Service Systems. G. M. and T. R. Simonson Company and E.D.A.C. A National Science Foundation report, May 1976.

Benkert, Donald. *Earthquake Protection of Vibration Isolated Equipment,* California Dynamics Corporation Sales Brochure, File 62, 1976.

Benuska, K. L., S. Aroni, and W. Schroll. "Elevator Earthquake Safety Control," *6th World Conference on Earthquake Engineering,* New Delhi, India, 1977.

"Earthquake Damage to General Telephone Company of California Facilities." General Telephone Company of California, Santa Monica, California. *San Fernando, California Earthquake of February 9, 1971.* L. M. Murphy, Ed. Volume 1, Part B. N.O.A.A., Washington, D.C., 1973.

"Earthquake Damage to Pacific Telephone and Telegraph Company Facilities." Pacific Telephone and Telegraph Company. *San Fernando, California Earthquake of February 9, 1971.* L. M. Murphy, Ed. Volume 1, Part B. N.O.A.A., Washington, D.C., 1973.

"Earthquake Damage to Southern California Edison Company Power Facilities." Southern California Edison Company. *San Fernando, California Earthquake of February 9, 1971.* L. M. Murphy, Ed. Volume 1, Part B. N.O.A.A., Washington, D.C., 1973.

Earthquake Resistant Regulations A World List 1973. Compiled by the International Association for Earthquake Engineering. Published by Gakujutsu Bunken Fukyu-Kai, Japan, April 1973.

Earthquake Resistant Regulations A World List 1973: Supplement 1976. Compiled by the International Association for Earthquake Engineering. Published by Gakujutsu Bunken Fukyu-Kai, Japan, 1976.

Eckel, E. B. "The Alaskan Earthquake: March 27, 1964, Lessons and Conclusions," *U.S.G.S. Prof. Paper 546,* 1970.

346   Earthquake Protection of Essential Building Equipment

Housner, George W. "Strong Ground Motion," *Earthquake Engineering*. Robert L. Wiegel, Ed., Prentice-Hall, Inc., Englewood Cliffs, N.J., 1970, Chapter 4.

IEEE Standard 344—1975, *IEEE Recommended Practices for Seismic Qualification of Class IE Equipment for Nuclear Power Generating Stations,* New York, 1975.

Iacopi, Robert. *Earthquake Country*. A Sunset Book, Lane Magazine and Book Company, Menlo Park, California, 1971.

Ibáñez, Paul, and W. E. Gundy. "Plant Engineering in Earthquake Country," *Proceedings of the Western Plant Engineering Conference and Show,* Anaheim, California, 1977.

Iwan, W. D. "Predicting the Earthquake Response of Resiliently Mounted Equipment with Motion Limiting Constraints," *6th World Conference on Earthquake Engineering,* New Delhi, India, 1977.

Jordan, Carl H. "Seismic Restraint of Equipment in Buildings," *Proceedings of the A.S.C.E. Annual Convention,* Volume 104, Number ST.5, 1978.

Kemper, Alfred M. *Architectural Handbook*. John Wiley & Sons, New York, 1979.

Leeds, David J. "The Design Earthquake," *Geology, Seismicity, and Environmental Impact*. Association of Engineering Geologists Special Publication. D. M. Moran, J. E. Slosson, R. O. Stone, and C. A. Yelverton, Eds. University Publishers, Los Angeles, California, October 1973.

Legget, Robert F. *Cities and Geology*. McGraw-Hill Book Company, New York, 1973.

Liu, S. C. "Earthquake Protection of Communication Facilities," *6th World Conference on Earthquake Engineering,* New Delhi, India, 1977.

Long Beach Municipal Building Code, California, 1973.

Los Angeles Building Code, California, 1975.

Mason, Norman J., and Patrick J. Lama. *Seismic Control Specifications for Floor Mounted Equipment*. Mason Industries, Inc. Sales Brochure SCS-100 Bulletin, Los Angeles, California, no date.

McCue, Gerald M., Ann Skaff, and John W. Boyce. *Architectural Design of Building Components for Earthquakes*. Sponsored by the National Science Foundation, Research Applied to National Needs Program, Washington, D.C., 1978.

McGavin, Gary L. *Seismic Qualification of Nonstructural Equipment in Essential Facilities,* unpublished Master of Architecture thesis, California State Polytechnic University, Pomona, 1978.

McGavin, Gary L. "An Examination of Aseismic Legislation for Nonstructural Components in Essential Facilities," *Proceedings of the 2nd U.S. National Conference on Earthquake Engineering,* Earthquake Engineering Research Institute, Stanford, California, 1979.

Merz, K. L. "Seismic Design & Qualification Procedures for Equipment Components of Lifeline Systems," *The Current State of Knowledge of Lifeline Earthquake Engineering,* Proceedings of the Technical Council on Lifeline Earthquake Engineering, A.S.C.E., U.C.L.A., 1977.

Miller, Richard K., and Stephen F. Felszeghy. *Engineering Features of the Santa Barbara Earthquake of August 13, 1978,* University of California Santa Barbara-ME-78-2 and Earthquake Engineering Research Institute, December 1978.

Muto, Kiyoshi, Masayuki Nagata, and Eiji Fukuzawa. "Earthquake Resistant Installation Device of Computers," *6th World Conference on Earthquake Engineering,* New Delhi, India, 1977.

Newmark, Nathan M., and E. Rosenblueth. *Fundamentals of Earthquake Engineering*. Prentice-Hall, Inc., Englewood Cliffs, N.J., 1971.

Recommended Lateral Force Requirements and Commentary. Seismology Committee Structural Engineers Association of California, San Francisco, 1975.

Reitherman, Robert K. "Some Notes on Nonstructural Features of the August 6, 1979 Coyote Lake Earthquake," *Earthquake Engineering Research Institute Newsletter,* November 1979.

Richter, Charles F. *Elementary Seismology.* W. H. Freeman and Company, San Francisco, 1958.

Roberts, Charles W. "Environmental Simulation—A Powerful Tool for the Product Designer," unpublished Wyle Laboratories paper, El Segundo, California, 1977.

Rule of General Application: Recommended Standards for Suspended Ceiling Assemblies, Los Angeles, California, 1974.

Seismic Design for Buildings. (Draft) Technical Manual No. 5-809-10, Departments of The Army, Air Force and Navy, August 1979.

Seismic Safety Design for Police and Fire Stations. A.I.A. Research Corp. and Public Technology, Inc., Washington D.C., 1977.

Shipway, George. "Seismic Seminar," unpublished Wyle Laboratories seminar text, El Segundo, California, 1974.

Specification 9618, Part F, Division F1, Section F1A, Article 27, City of Los Angeles Department of Water and Power, issued in connection with Energy Control Project, no date.

State of California, California Administrative Code, Title 24, Building Standards, Part 3, Basic Electrical Regulations, Sacramento, California, 1974.

State of California, California Administrative Code, Title 24 Building Standards, Part 4, Basic Mechanical Regulations, Sacramento, California, 1971.

State of California, California Administrative Code, Title 24 Building Standards, Part 5, Basic Plumbing Regulations, Sacramento, California, 1971.

State of California, California Administrative Code, Title 24 Building Standards, Part 6, Special Building Regulations, Division T17, State Department of Public Health, Sacramento, California, 1974.

State of California, California Administrative Code, Title 24 Building Standards, Part 7, Elevator Safety Regulations, Sacramento, California, 1977.

Steinbrugge, K. V., and H. J. Degenkolb. "Meeting the Earthquake Challenge: California's New Laws." *Civil Engineering,* American Society of Civil Engineers, February 1975.

Stone, Marraccini, and Patterson. *Study to Establish Seismic Protection Provisions for Furniture, Equipment and Supplies for VA Hospitals,* Research Staff Office of Construction, Veterans Administration, Washington, D.C., 1976.

Sugimoto, Yoneo, and Yuji Sato. "Telecommunications Equipment Seismic Effect Study," *6th World Conference on Earthquake Engineering,* New Delhi, India, 1977.

Summary of Veterans Administration Earthquake Engineering Program. Veterans Administration, Washington, D.C., March 1977.

Summer Seismic Institute for Architectural Faculty. A.I.A. Research Corporation. Sponsored by the National Science Foundation, Research Applied to National Needs Program, Stanford University, 1977.

Tentative Provisions for the Development of Seismic Regulations for Buildings, prepared by Applied Technology Council, ATC 3-06, National Bureau of Standards, NBS 510, National Science Foundation, NSF 78-8, Washington, D.C., 1978.

Uniform Building Code. International Conference of Building Officials, Whittier, California, 1979.

Urban Geology: Master Plan for California, California Division of Mines and Geology, Bulletin 198, 1973.

VA Construction Standard CD-54; H-08-3. Veterans Administration, Washington, D.C., August 1975.

VA *Construction Standard CD-55; H-08-3*. Veterans Administration, Washington, D.C., November 1973.

Wong, Patrick P. "Earthquake Effects on Power System Facilities of the City of Los Angeles." *San Fernando, California Earthquake of February 9, 1971*. L. M. Murphy, Ed. Volume 1, Part B. N.O.A.A., Washington, D.C., 1973.

Yancey, C. W. C., and A. A. Camacho. Aseismic Design of Building Service Systems. The State-of-the-Art, National Bureau of Standards Technical Note 970, Washington, D.C., 1978.

Yanev, Peter I. *Peace of Mind in Earthquake Country*. Chronicle Books, San Francisco, California, 1974.

Yuceoglu, U. *Seismic Design of Equipment Supports and Connections in Industrial Installations—A Brief Summary*. ASME paper 80-C2/PVP-73, New York, N.Y., 1980.

Appendix 1. Codes and Specifications

DOMESTIC SEISMIC CODES AND SPECIFICATIONS REVIEWED IN APPENDIX 1

- Yancey and Camacho, 1978, Code Identification Tables
  - Tables
  - References
- International Conference of Building Officials *Uniform Building Code*, 1979 Edition (UBC 1979)
  - Excerpts
  - General Discussion of UBC 1979
  - Proposed Changes to the *Uniform Building Code*
  - Discussion of Proposed Changes to the *Uniform Building Code*
- Structural Engineers Association of California "Recommended Lateral Force Requirements 1975"
- Applied Technology Council 3-06 "Tentative Provisions for the Development of Seismic Regulations for Buildings"
  - Excerpts
  - General Discussion of ATC 3-06
- *State of California, California Administrative Code,* Title 24, Building Standards, Part 3 (CAC 24-3), "Basic Electrical Regulations"
- *State of California, California Administrative Code,* Title 24, Building Standards, Part 4 (CAC 24-4), "Basic Mechanical Regulations"
- *State of California, California Administrative Code,* Title 24, Building Standards, Part 5 (CAC 24-5), "Basic Plumbing Regulations"
- *State of California, California Administrative Code,* Title 24, Building Standards, Part 6 (CAC 24-6), "Special Building Regulations"
- *State of California, California Administrative Code,* Title 24, Building Standards, Part 7 (CAC 24-7), "Elevator Safety Regulations"
  - Excerpts
  - General Discussion of CAC 24-7
- *Earthquake Resistant Design Requirements for VA Hospital Facilities,* Handbook H-08-8, Veterans Administration (VA H-08-8)

349

- *Post-Earthquake Emergency Utility Services and Access Facilities, Veterans Administration Construction Standard CD-54 (VA CD-54)*
- *Earthquake-Resistive Design of Nonstructural Elements of Buildings, Veterans Administration Construction Standard CD-55 (VA CD-55)*
- City of Los Angeles: *Los Angeles Municipal Code*
- City of Los Angeles: *Recommended Standards for Suspended Ceiling Assemblies*
- City of Los Angeles Department of Water and Power Specification 9618, Part F, Division F1, Section F1A, Article 27, Issued in Connection with Energy Control Project
  - Excerpt

FOREIGN SEISMIC CODES REVIEWED IN APPENDIX 1

- Bulgaria: *Bulgarian Code for Building in Earthquake Regions,* 1964
- Canada: *National Building Code of Canada 1975*
- El Salvador: *Regulations for Seismic Design, Republic of El Salvador, C.A., 1966*
- France: *Building Regulations for Seismic Areas 1967*
- Israel: *Proposed Israel Standard Loads in Buildings: Earthquakes*
- New Zealand: *New Zealand Standard, NZS 4203:* 1976
- Peru: *Peruvian Standards for Antiseismic Design, 1968*
- Rumania: *Earthquake Regulations Rumanian People's Republic*
- Venezuela: *Provisional Standard for Earthquake-Resistant Structure,* 1967
- Concluding Remarks on Codes and Standards

DOMESTIC AND FOREIGN SEISMIC CODES AND SPECIFICATIONS

The continued operation of essential facilities (e.g., hospitals, police stations, communication centers, and emergency operating centers) is necessary to serve the public after a severe earthquake. Recent improvements in earthquake regulations and codes have resulted in buildings that are more likely to remain structurally intact after an earthquake. However, many types of equipment located within these facilities and required for the successful operation of the facility have not received adequate attention in existing building codes. Experience such as that obtained in the 1964 Alaskan earthquake and the 1971 San Fernando earthquake has shown that the equipment items required for the operation of the facility are commonly unable to perform their functions in the immediate earthquake aftermath. Equipment failure is a serious problem, and in many cases, human life depends on the continued operation of the equipment. Codes that are currently being enforced consistently ignore the dynamic nature of equipment as well as the operability requirements of specialized components such as communication systems and life support equipment.

The purpose of this appendix is to examine and comment on some seismic building codes that deal with equipment. Many of the codes reviewed are virtual repeats of "parent codes," while others, such as the Applied Technology Council recommendations (not yet enacted), have taken bold new steps to deal with equipment that will certainly increase the chance for equipment survival during and after a severe earthquake. Codes such as the *State of California, California Administrative Code,* Title 24, Building Standards, Part 7, "Elevator Safety Regulations" address the seismic environment in detail, but seriously restrict the functional aspects of essential facilities by requiring automatic elevator shutdown followed by detailed inspection sequences in the event of strong building motion. Aseismic qualification procedures that would not require elevator shutdown except where hoistway collisions are imminent are now available to the industry.

This examination of equipment earthquake legislation presents some of the various aspects of applicable codes, both positive and negative, that affect the functional capabilities of essential facilities. It will also provide the reader with a base for comparison and understanding when dealing with equipment design and installation, especially in essential facilities.

Building codes appear to have originated at least as early as Hammurabi's time. Strict earthquake design regulations, however, did not appear until early in the twentieth century. Significant progress was made between the 1930s and the early 1970s for structurally oriented seismic building codes. Then, in a time span of 7 years, two damaging earthquakes occurred within major populated areas of the United States (Anchorage and San Fernando). Detailed examinations immediately after these events led to new areas of concern for earthquake design. It was found that buildings whose structures were left relatively intact had, in many cases, suffered severe damage to nonstructural components and building equipment. The observations included damage to curtain walls, facades, partition walls, emergency power supplies, library stacks, communication equipment, and hospital equipment.

Building component damage is of special concern where the building has a specific or essential function to perform during or immediately after an earthquake. The first major design consideration for nonstructural components appeared in the Structural Engineers Association of California (SEAOC) *Recommended Lateral Force Requirements—1975.* These recommendations have since been adopted by the *Uniform Building Code* (UBC 1976 and UBC 1979) and various California administrative codes with minor additions and/or revisions. The SEAOC recommendations also contain the first mention of essential facilities. This "parent code" requires the operability of essential facilities immediately after an earthquake, but it does not suggest a suitable vehicle for implementation of the operability requirement. For the most part, all the codes ignore the dynamic nature of the specific applications of the codes by simply requiring a static coefficient approach for equipment installation. This approach is often suitable for components only needing to demonstrate structural integrity such as base anchorage. Critical equipment (life support, etc.) is, however, commonly more complex and often must physically operate both during and after a severe earthquake. This makes an adequate qualification procedure more complex and requires an approach different from the static coefficient method, such as a dynamic analysis and/or a seismic test. Only one proposed code—the Applied Technology Council ATC 3-06 *Tentative Provisions for the Development of Seismic Regulations for Buildings*—currently suggests such an approach.

The purpose of reviewing codes and specifications that address equipment items is to place the codes in a perspective for comparison so that the design team can better implement them. The discussion is limited only to the equipment portions of the code rather than all the seismic conditions that the code addresses. Codes that are obvious duplications of other codes are so noted and are only discussed if they have any distinguishing features.

YANCEY AND CAMACHO, 1978, CODE IDENTIFICATION TABLES

Tables

Tables A1.1–A1.5 are adapted from Yancey and Camacho's report on *Aseismic Design of Building Service Systems,* National Bureau of Standards Technical Note 970, September 1978, Tables 14–18. These tables present various equipment systems and equipment items and any corresponding aseismic design codes and/or specifications. All the design guide references to ATC 3-06 have been altered from Yancey and Camacho's original report. Their tables reflected the draft of ATC 3, which has since been published.

TABLE A1.1.   Identification of Available Codes and Design Guides—Fire Protection System

Essential Service System	Name of Code or Standard [Yancey and Camacho Reference No.]	Design Guides [Yancey and Camacho Reference No.]
FIRE PROTECTION		
Sprinkler system	*National Fire Code,* Vol. 13, Chap. 3 [26]; *California Administrative Code,* Title 17 [12]; *San Francisco Building Code* [30]	*National Fire Code,* Vol. 13, Appendix 13 [26]; ATC-3, Chap. 8 [6]
Risers	Same as above	Same as above
Distribution mains	Same as above	Same as above
Valves	Same as above	Same as above

TABLE A1.1.

Essential Service System	Name of Code or Standard [Yancey and Camacho Reference No.]	Design Guides [Yancey and Camacho Reference No.]
Branch pipes	Same as above	Same as above
Sprinkler heads and controls	—	—
Support hangers, bracing, and controls	Same as above	Same as above
Support hangers, bracing, and clamps	Same as above	Same as above
Standpipes	*National Fire Code*, Vol. 13, Chap. 3 [26]; *California Administrative Code*, Title 17	*National Fire Code*, Vol. 13, Appendix 13 [26]; ATC-3, Chap. 8 [6]
Mains	Same as above	Same as above
Risers	Same as above	Same as above
Clamps and hangers	Same as above	Same as above
Pumps		
Main unit	Not applicable	Not applicable
Pipe connections		
Supports	*San Francisco Building Code* [30]; *California Administrative Code*, Title 17 [12]; *Tri-Services Manual* [36]; UBC [37]; *Los Angeles Building Code* [21]	ATC-3, Chap. 8 [6]
Pressure tanks		
Tank		
Supports	*San Francisco Building Code* [30]; *California Administrative Code*, Title 17 [12]; *Tri-Services Manual* [36]; UBC [37]; *Los Angeles Building Code* [21]	ATC-3, Chap. 8 [6]
Suction tanks		
Tank		
Supports	*San Francisco Building Code* [30]; *California Administrative Code*, Title 17 [12]; *Tri-Services Manual* [36]; UBC [37]; *Los Angeles Building Code* [21]	ATC-3, Chap. 8 [6]

Source: Yancey, C. W. C. and A. A. Camacho, *Aseismic Design of Building Service Systems: The State-of-the-Art*, NBS Technical Note 970, U.S. Department of Commerce/National Bureau of Standards, issued September 1978.

TABLE A1.2. Identification of Available Codes and Design Guides—Emergency Power System

Essential Service System	Name of Code or Standard [Yancey and Camacho Reference No.]	Design Guides [Yancey and Camacho Reference No.]
EMERGENCY POWER		
Motor–generator set		
Motor and generator	NA*a*	NA
Radiator	NA	NA
Piping	NA	NA
Controls	NA	NA
Fuel piping	NA	ATC-3, Chap. 8 [6]
Starting batteries	NA	
Mufflers		
Supports	VA Handbook H08-8 [16]; *California Adminis-trative Code,* Title 17 [12]; *Tri-Services Manual* [36]; *San Francisco Building Code* [30]	ATC-3, Chap. 8 [6]
Transformers		ATC-3, Chap. 8 [6]
Main unit		Same as above
Wiring connections		Same as above
Supports	VA Handbook H08-8 [16]; *California Adminis-trative Code,* Title 17 [12]; *Tri-Services Manual* [36]; *San Francisco Building Code* [30]	
Switchgear		ATC-3, Chap. 8 [6]
Main unit		
Conduits		Same as above
Supports	*San Francisco Building Code* [30]; Handbook H08-8 [16]; *California Administrative Code,* Title 17 [12]; *Tri-Services Manual* [36]	Same as above
Panelboards		ATC-3, Chap. 8 [6]
Housing		Same as above
Conduits		Same as above
Supports	*San Francisco Building Code* [30]; Handbook H08-8 [16]; *California Administrative Code,* Title 17 [12]; *Tri-Services Manual* [36]	

TABLE A1.2. (Continued)

Essential Service System	Name of Code or Standard [Yancey and Camacho Reference No.]	Design Guides [Yancey and Camacho Reference No.]
Electrical distribution network	*San Francisco Building Code* [30]; VA Handbook H08-8 [16]; *California Administrative Code*, Title 17 [12]	ATC-3, Chap. 8 [6]
Bus ducts	Same as above	Same as above
Feeders	Same as above	Same as above
Connectors	Same as above	Same as above
Supports	Same as above	Same as above
Lighting		
Lighting fixtures	*Tri-Services Manual* [36]; *California Administrative Code*, Title 17 [12]; RGA 12-69 [20]; VA Handbook H08-8 [16]	ATC-3, Chap. 8 [6]
Recessed	*Tri-Services Manual* [36]; *California Administrative Code*, Title 17 [12]; VA Handbook H08-8 [16]	Same as above
Surface-mounted	*Tri-Services Manual* [36], VA Handbook H08-8 [16]; *California Administrative Code*, Title 17 [12]	Same as above
Stem and chain suspended	*Tri-Services Manual* [36], RGA 12-69 [20]; VA Handbook H08-8 [16]; *California Administrative Code*, Title 17 [12]	Same as above

Source: Yancey, C. W. C. and A. A. Camacho, *Aseismic Design of Building Service Systems: The State-of-the-Art,* NBS Technical Note 970, U.S. Department of Commerce/National Bureau of Standards, issued September 1978.
[a] Not applicable.

TABLE A1.3. Identification of Available Codes and Design Guides—Sanitation and Water Supply System

Essential Service System	Name of Code or Standard [Yancey and Camacho Reference No.]	Design Guides [Yancey and Camacho Reference No.]
SANITATION AND WATER SUPPLY		
Pumps and motors		
Main unit	NA[a]	NA
Pipe connections	NA	NA
Supports	VA Handbook H08-8 [16]; *California Administrative Code,* Title 17 [12]; *San Francisco Building Code* [30]; *Tri-Services Manual* [36]	ATC-3, Chap. 8 [6]
Hot and cold water storage tanks		
Tank body	NA	NA
Pipe connections	NA	NA
Supports	VA Handbook H08-8 [16]; UBC [37]; *California Administrative Code,* Title 17 [30]; *San Francisco Building Code* [30]; *Tri-Services Manual* [36]	ATC-3, Chap. 8 [6]
Piping (air, steam, vacuum, gas)	VA Handbook H08-8 [16]; *Tri-Services Manual* [36]	ATC-3, Chap. 8 [6]; *Tri-Services Manual,* Appendix H [36]
Pipes	Same as above	
Fittings	Same as above	
Supports	Same as above	
Water heaters		
Heater body	NA	NA
Pipe connections	NA	NA
Supports	VA Handbook H08-8 [16]; *California Administrative Code,* Title 17 [12]	ATC-3, Chap. 8 [6]
Plumbing fixtures		

Source: Yancey, C. W. C. and A. A. Camacho, *Aseismic Design of Building Service Systems: The State-of-the-Art,* NBS Technical Note 970, U.S. Department of Commerce/National Bureau of Standards, issued September 1978.
[a] Not applicable.

TABLE A1.4. Identification of Available Codes and Design Guides—Environmental Control System

Essential Service System	Name of Code or Standard [Yancey and Camacho Reference No.]	Design Guides [Yancey and Camacho Reference No.]
ENVIRONMENTAL CONTROL		
Compressors (air, medi-cal, refrigeration)		
Main unit	NA	NA
Pipe connections		
Supports	VA Handbook H08-8 [16]; *California Adminis-trative Code*, Title 17 [12]; *Tri-Services Man-ual* [36]; *San Francisco Building Code* [30]	ATC-3, Chap. 8 [6]
Fans (air supply, exhaust)		
Main unit		
Supports	VA Handbook H08-8 [16]; *California Adminis-trative Code*, Title 17 [12]; *Tri-Services Man-ual* [36]; *San Francisco Building Code* [30]	ATC-3 Chap. 8 [6]
Chillers		
Main unit		
Pipe connections		
Supports	VA Handbook H08 8 [16]; *California Adminis-trative Code*, Title 17 [12]; *Tri-Services Man-ual* [36]; *San Francisco Building Code* [26]	
Boilers		
Main unit		
Pipe connections		
Supports	VA Handbook H08-8 [16]; *California Administra-tive Code*, Title 17 [12]; *Tri-Services Manual* [36]; *San Francisco Building Code* [30]	ATC-3, Chap. 8 [6]
Duct network		
Main distribution ducts	*Tri-Services Manual* [36]	ATC-3 Chap. 8 [6]
Branch distribution ducts	*Tri-Services Manual* [36]	ATC-3 Chap. 8 [6]

Essential Service System	Name of Code or Standard [Yancey and Camacho Reference No.]	Design Guides [Yancey and Camacho Reference No.]
Heat exchangers		
Main unit		
Pipe connections		
Supports	VA Handbook H08-8 [16]; *California Adminis- trative Code,* Title 17 [12]; *Tri-Services Man- ual* [36]; *San Francisco Building Code* [30]	ATC-3 Chap. 8 [6]
Chimneys, flues, and vents	UBC [37]; *California Ad- ministrative Code,* Title 17 [12]; *Los Angeles Building Code* [20], *San Francisco Building Code* [30]	ATC-3 Chap. 8 [6]
HVAC and fuel piping	*Tri-Services Manual* [36]; VA Handbook H08-8 [16]; *California Admin- istrative Code,* Title 17 [12]	ATC-3 Chap. 8 [6]
Pipes	Same as above	Same as above
Fittings	Same as above	Same as above
Supports	Same as above	Same as above
Pumps		
Main unit		
Pipe connections		
Supports	*Tri-Services Manual* [36]; VA Handbook H08-8 [16]; *California Admin- istrative Code,* Title 17 [12]	ATC-3, Chap. 8 [6]
Condensers		
Main unit		
Pipe connections		
Supports	*Tri-Services Manual* [36]; VA Handbook H08-8 [16]; *California Admin- istrative Code,* Title 17 [12]	ATC-3 Chap. 8 [6]

Source: Yancey, C. W. C. and A. A. Camacho, *Aseismic Design of Building Service Systems: The State-of-the-Art,* NBS Technical Note 970, U.S. Department of Commerce/National Bureau of Standards, issued September 1978.

TABLE A1.5. Identification of Available Codes and Design Guides—General Services System

Essential Service System	Name of Code or Standard [Yancey and Camacho Reference No.]	Design Guides [Yancey and Camacho Reference No.]
GENERAL SERVICES		
People movers		
Elevators—traction type	*San Francisco Building Code* [30]; *California Administrative Code,* Title 17 [12]; VA Handbook H08-8 [16]	
Guide rails	Same as above	
Motor–generators		
Counterweights	Same as above	
Control panels		
Cars	Same as above	
Support system	Same as above	
Elevators—hydraulic type	Same as above	
Escalators		
Machine and drive		
Controllers		
Trusses and tracks		
Communication system	*California Administrative Code,* Title 17 [12]; VA Handbook H08-8 [16]	ATC-3, Chap. 8 [6]
Intercom/pa system	Same as above	Same as above
Telephone equipment	Same as above	Same as above
Switchboards	Same as above	Same as above

Source: Yancey, C. W. C. and A. A. Camacho, *Aseismic Design of Building Service Systems: The State-of-the-Art,* NBS Technical Note 970, U.S. Department of Commerce/National Bureau of Standards, issued September 1978.

REFERENCES NOTED IN TABLES A1.1–A1.5

6. Applied Technology Council, *Tentative Provisions for the Development of Seismic Regulations for Buildings,* ATC 3-06, Chapter 8, "*Architectural, Mechanical and Electrical Components and Systems,*" NBS Special Publication 510, NSF Publication 78-8.

12. *State of California, California Administrative Code,* Title 17, Public Health, Chapter 8, "Safety of Construction of Hospitals," State of California, 1973.

16. *Earthquake Resistant Design Requirements for VA Hospital Facilities,* Handbook H08-8, Office of Construction, Veterans Administration, June 1973 (4/74).

20. Los Angeles, City of, Department of Building and Safety, *Rule of General Application*, RGA 12-69, "Standard for Lighting Fixture Supports," 1979.

21. *Los Angeles Building Code*, Division 23, "Loads and General Design," 1973 edition and 1973 and 1974 amendments.

26. *National Fire Code*, Vol. 13, "Installation of Sprinkler Systems," National Fire Protection Association.

30. *San Francisco Building Code*, Article 23, 1975 edition.

36. *Tri-Services Manual*, Army TM 5-809-10, Navy NAV FAC P-355, Air Force AFM 88-3, "Seismic Design of Buildings," Department of the Army, the Navy, and the Air Force, April 1973.

37. *Uniform Building Code*, 1973 edition, International Conference of Building Officials.

INTERNATIONAL CONFERENCE OF BUILDING OFFICIALS *UNIFORM BUILDING CODE*, 1979 EDITION (UBC 1979)

Excerpts

All *Uniform Building Code* excerpts are reproduced from the 1979 edition of the *Uniform Building Code*, copyrighted 1979, with permission of the publisher, the International Conference of Building Officials.

The shaded margin indicates portions of the code that were added or amended in 1979.

. . . . . .

2312(g) Lateral Force on Elements of Structures and Nonstructural Components. Parts or portions of structures, nonstructural components and their anchorage to the main structural system shall be designed for lateral forces in accordance with the following formula:

$$F_p = ZIC_pW_p \qquad (12\text{-}8)$$

The values of C_p are set forth in Table No. 23-J (Table A1.6 in this book). The value of the I coefficient shall be the value used for the building.
EXCEPTIONS: 1. The value of I for panel connectors shall be as given in Section 2312 (j) 3 C.

2. The value of I for anchorage of machinery and equipment required for life safety systems shall be 1.5.

The distribution of these forces shall be according to the gravity loads pertaining thereto.

For applicable forces on diaphragms and connections for exterior panels, refer to Sections 2312 (j) 2 D and 2312 (j) 3 C.

. . . . . .

(k) Essential Facilities. Essential facilities are those structures or buildings which must be safe and usable for emergency purposes after an earthquake in order to preserve the health and safety of the general public. Such facilities shall include but not be limited to:

1. Hospitals and other medical facilities having surgery or emergency treatment areas.
2. Fire and police stations.
3. Municipal government disaster operation and communication centers deemed to be vital in emergencies.

   The design and detailing of equipment which must remain in place and be functional following a major earthquake shall be based upon the requirements of Section 2312 (g) and Table No. 23-J. In addition, their design and detailing shall consider effects induced by structure drifts of not less than $(2.0/K)$ times the story drift caused by required seismic forces nor less than the story drift caused by wind. Special consideration shall also be given to relative movements at separation joints.

   . . . . . .

General Discussion of UBC 1979

The UBC 1979, Paragraph 2312, Section (k) principally defines essential facilities as those structures or buildings that must be usable for emergency purposes after an earthquake. Examples are hospitals, fire and police stations, communication centers, and municipal emergency operating centers. The essential facility definition was a new addition to the UBC 1976; it is concise and specific and is not difficult to interpret.

Table 23-J (table A1.6 in this book) of the UBC 1979 shows which items are covered by the nonstructural component clause, the direction from which the seismic force must be applied, and the numerical coefficient (C_p) for application to UBC 1979 Equation 12-8 $(F_p = ZIC_pW_p)$. Table 23-K (Table A1.7 in this book) supplies the importance coefficient for various facility types.

Table 23-J, numbers 1, 2, 3, and 6 do not apply to this discussion. Table 23-J, number 4 (a–c) does not require examination of owner supplied equipment that is necessary for life safety systems or the continued operation of essential facilities. Footnote d removes the burden of installation detailing from the architect or engineer by stating:

> The design of the equipment and machinery and their anchorage is an integral part of the design and specification of such equipment and machinery.

This note totally defeats the intent of the definition of essential facilities and removes the requirement for the architect or engineer to adequately detail building equipment.

The equivalent static coefficient approach (Equation 12-8) taken by UBC 1979 works relatively well for equipment that only needs to remain anchored. More complex equipment, however, is more likely to fail operationally rather than by mere anchorage alone and is not adequately qualified by the equivalent static coefficient approach. Dynamic considerations should be provided by future editions of the UBC. Such dynamic considerations

TABLE A1.6. Horizontal Force Factor C_p for Elements of Structures and Nonstructural Components (Table 23-J)

Part or Portion of Buildings	Direction of Horizontal Force	Value of $C_p{}^a$
1. Exterior bearing and nonbearing walls, interior bearing walls and partitions, interior nonbearing walls and partitions—see also Section 2312 (j) 3 C. Masonry or concrete fences over 6 feet high	Normal to flat surface	0.3^b
2. Cantilever elements: a. Parapets	Normal to flat surfaces	0.8
b. Chimneys or stacks	Any direction	
3. Exterior and interior ornamentations and appendages	Any direction	0.8
4. When connected to, part of, or housed within a building: a. Penthouses, anchorage and supports for chimneys and stacks and tanks, including contents		
b. Storage racks with upper storage level at more than 8 feet in height, plus contents	Any direction	$0.3^{c,d}$
c. All equipment or machinery		
5. Suspended ceiling framing systems (applies to Seismic Zones Nos. 2, 3, and 4 only)	Any direction	0.3^e
6. Connections for prefabricated structural elements other than walls, with force applied at center of gravity of assembly	Any direction	0.3^f

$^a C_p$ for elements laterally self-supported only at the ground level may be two-thirds of value shown.

^{b} See also Section 2309 (b) for minimum load and deflection criteria for interior partitions.

$^c W_p$ for storage racks shall be the weight of the racks plus contents. The value of C_p for racks over two storage support levels in height shall be 0.24 for the levels below the top two levels. In lieu of the tabulated values steel storage racks may be designed in accordance with UBC Standard No. 27-11.

Tables A1.6-A1.7 reproduced from the 1979 Edition of the *Uniform Building Code*, Copyright 1979, with permission of the publisher. The International Conference of Building Officials.

Where a number of storage rack units are interconnected so that there are a minimum of four vertical elements in each direction on each column line designed to resist horizontal forces, the design coefficients may be as for a building with K values from Table No. 23-1, $CS = 0.2$ for use in the formula $V = ZIKCSW$ and W equal to the total dead load plus 50 percent of the rack-rated capacity. Where the design and rack configurations are in accordance with this paragraph, the design provisions in UBC Standard No. 27-11 do not apply.

[d]For flexible and flexibly mounted equipment and machinery, the appropriate values of C_p shall be determined with consideration given to both the dynamic properties of the equipment and machinery and to the building or structure in which it is placed but shall be not less than the listed values. The design of the equipment and machinery and their anchorage is an integral part of the design and specification of such equipment and machinery.

For essential facilities and life safety systems, the design and detailing of equipment which must remain in place and be functional following a major earthquake shall consider drifts in accordance with Section 2312 (k).

[e]Ceiling weight shall include all light fixtures and other equipment which is laterally supported by the ceiling. For purposes of determining the lateral force, a ceiling weight of not less than 4 pounds per square foot shall be used.

[f]The force shall be resisted by positive anchorage and not by friction.

TABLE A1.7. Values for Occupancy Importance
Factor I. (Table 23-K)

Type of Occupancy	I
Essential facilities[a]	1.5
Any building where the primary occupancy is for assembly use for more than 300 persons (in one room)	1.25
All others	1.0

Reproduced from the 1979 edition of the *Uniform Building Code*, Copyright 1979, with permission of the publisher, The International Conference of Building Officials.

[a]See Section 2312 (k) for definition and additional requirements for essential facilities.

would require dynamic analysis for some equipment and actual seismic testing for other equipment items depending on their complexity and function. Currently, mention of dynamic properties by UBC 1979 is only made for flexible or flexibly mounted equipment.

The use of UBC 1979 Equation 12-8 is fairly straightforward. The equation is

$$F_p = ZIC_pW_p$$

where F_p = lateral force on part of the structure and in the direction under consideration

Z = the seismic zone coefficient from Figure 1, 2, or 3 (Figures A1.1, A1.2, and A1.3 in this book)

Zone 0, $Z = 0$
Zone 1, $Z = \frac{3}{16}$
Zone 2, $Z = \frac{3}{8}$
Zone 3, $Z = \frac{3}{4}$
Zone 4, $Z = 1$

I = occupancy importance factor from Table 23-K
C_p = numerical coefficient from Section 2312 (g) and Table 23-J
W_p = weight of the equipment

For an example application, consider the situation where a cold water pump is required for installation in the basement of a hospital. The pump weighs 500 pounds and is provided with four $\frac{3}{4}$-inch diameter anchorage points 18 inches on center in the base plate by the manufacturer. Anchorage is fixed and vibration isolation is not used. The example hospital is located in Los Angeles, California. With this information, the problem is approached as follows:

$Z = 1$ (zone 4 for Los Angeles, California)
$I = 1.5$ (hospitals are essential facilities)
$C_p = 0.3$ (Table 23-J, number 4c)
$W_p = 500$ pounds

Working the problem with the variables leads to:

$F_p = ZIC_pW_p$
   $= (1)(1.5)(0.3)(500)$
   $= 225$ pounds

Each anchor point must, therefore, be able to resist one-quarter of the total force or 56.25 pounds of horizontal shear. Standard $\frac{3}{4}$-inch diameter anchor bolts more than meet this requirement. Calculations for anchor pullout should also be performed. By UBC 1979 standards, the pump is qualified at this point. Other problems that deserve attention are not addressed by this code.

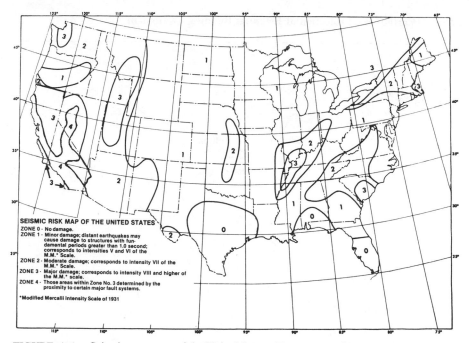

FIGURE A.1. Seismic zone map of the United States. For areas outside the United States see Appendix Chapter 23 of the UBC 79. Reproduced from the 1979 edition of the *Uniform Building Code*, Copyright 1979, with permission of the publisher, The International Conference of Building Officials.

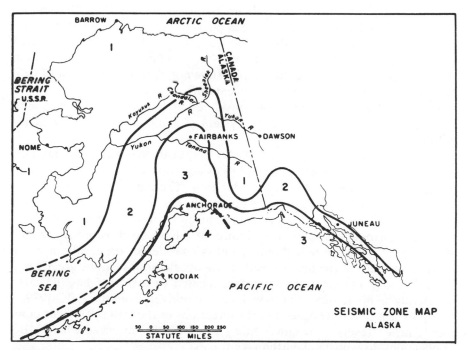

FIGURE A.2. Seismic zone map of Alaska. Reproduced from the 1979 edition of the *Uniform Building Code*, Copyright 1979, with permission of the publisher, The International Conference of Building Officials.

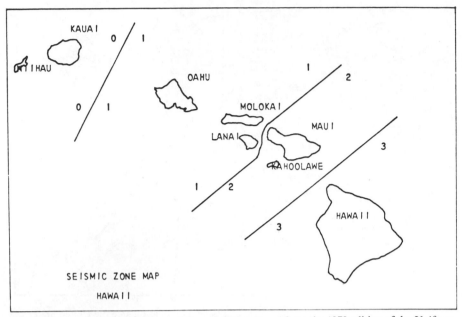

KAUA I

0 / 1

I HAU

OAHU

0 / 1

1 / 2

MOLOKA I

LANA I

MAU I

3

KAHOOLAWE

1 / 2

HAWA I I

3

SEISMIC ZONE MAP

HAWA I I

FIGURE A.3. Seismic zone map of Hawaii. Reproduced from the 1979 edition of the *Uniform Building Code*, Copyright 1979, with permission of the publisher, The International Conference of Building Officials.

If rigid inlet and outlet water lines are used, the UBC 1979 does not consider flange loads. These loads can possibly add to the lateral load and thereby exceed the rating of the anchorage provided by the code calculated value. If rigid lines are used, the pump is likely to fail because of ruptures in the line. Example after example can be shown illustrating this point of deficiency in the code approach to qualification.

A piece of equipment should be considered by building codes not as individually but rather as a part of the system of equipment items to which it belongs.

An examination of the product $(Z)(I)(C_p)$ reveals that it is equal basically to the acceleration in the classical physics equation where force is equal to the mass of an object times its acceleration. The equivalent accelerations that can be found in the UBC 1979 do not approach reality if one considers the accelerations that earthquakes produce. Ground accelerations from moderate earthquakes such as El Centro 1979 have been recorded at much higher levels. When building equipment is considered, especially in multistoried buildings at the upper levels, these accelerations can be greatly amplified. Strong motion instrumentation located in buildings has indicated accelerations many times that which can be derived from the UBC 1979.

Equipment that is designed per the directions of the UBC 1979 does commonly have "reserve strength." Many cases can be observed where UBC applications do withstand earthquake forces.

The *Uniform Building Code* is, to date, not a comprehensive seismic code with respect to building equipment. It does, however, provide some of the basics for an earthquake qualification program. It is a growing code, and as more design professionals become aware of the importance of building equipment survivability for safety and economic reasons, the UBC will certainly adapt to the profession's awareness.

Proposed Changes to the *Uniform Building Code*

Proposing code change is a relatively simple task. Having the proposal incorporated, however, is another matter. The International Conference of Building Officials published their most recent version of the *Uniform Building Code* in 1979. The State of California is just now holding hearings to incorporate portions of the UBC 1976 into the *California Administrative Code* (CAC) for public schools, and some municipalities still adhere to the UBC 1973.

Even though the structures remained intact, recent earthquakes have closed a number of essential facilities because of equipment failures. Photographic examples of these facilities are included at the close of Chapter 3.

The importance of equipment operation and anchorage was not considered by early codes and is not yet stringent enough. The 1971 San Fernando failures sparked an immediate upgrading of both the UBC (1973 edition) and the CAC (1974 revisions). Strong motion records and observations of failures in past earthquakes indicated that the upgraded 1973 and 1976 lateral load factors (static equivalents) were still not enough. The UBC upgraded its lateral load requirements for nonstructural equipment and added the first mention of essential facilities in 1976. The CAC 1974 was slightly more definitive than the UBC 1973, but less so than the UBC 1976 and UBC 1979. Studies by independent companies such as Mason Industries and Wyle Laboratories have shown that the static lateral load factors still do not represent the real environment experienced by many pieces of nonstructural equipment during an earthquake. This has been shown repeatedly by Wyle Laboratories' seismic testing programs for the nuclear power industry and the Alaska Pipeline.

Strong motion recordings have shown that dynamic loads at component and subcomponent locations during past earthquakes have commonly exceeded one full g. This is enough of a load to "throw" an unattached object, whatever its weight. In special cases, subcomponents may experience as much as $20g$ because of the many variables associated with equipment installation and response. Designing for high loads, such as these, is not required by any of the current seismic codes. The UBC 1979 only considers required equipment from a static viewpoint. Until this situation is corrected, essential facilities can be expected to fail during future earthquakes.

The UBC 1976 introduced the essential facility phrase to seismic building codes. It also stated that essential facilities must remain operational after an earthquake. Mention is not made concerning operation during the earthquake for life support equipment or other equipment required for the facility operation. Earthquakes generally last less than 1 minute as far as severe ground shaking is concerned. That small time frame, however, can result in a life or death situation for many individuals. Patients on dialysis machines, for example, are not likely to survive if air is suddenly and accidentally introduced into the deionized water supply. It is the purpose of the author to propose additions to the UBC that will clarify the operational requirement as well as the necessary dynamic considerations for adequate seismic qualification legislation.

Since most local legislation dealing with aseismic construction arises from the *Uniform Building Code,* the code proposals in this section are directed toward that publication alone. Future application of these proposals is not guaranteed and they may never be incorporated.

The imposed additional costs upon building owners will most certainly be a matter of concern. The potential gains received from the proposed changes will have to be weighed against the potential losses and costs arising from an earthquake. Adequate aseismic design can save money and lives in the long run. Much of the equipment found in an essential facility, for instance, is rather costly. If it fails, the facility runs the risk of failure in its designated task. If the equipment failure is bad enough, it may require replacement which often results in a cost higher than that of an initial seismic qualification program. When specific life support items are allowed to fail, individuals die needlessly. It is impossible to attach an economic significance to this loss of human life. Future statistic and economic studies, therefore, are needed to justify the proposals included in this section for acceptance to the *Uniform Building Code*.

This first proposal directs itself toward newly constructed essential facilities only. The capability of existing facilities to survive an earthquake is a problem that should not be ignored, but it has not been addressed in depth by this proposal. However, it is felt that this is a valid topic for future research programs. Backfitting programs are discussed in Chapter 3. The need for these programs should not be ignored, and walk-through seismic safety tours should be conducted by personnel of existing facilities to determine where backfitting is needed.

Presented below are the proposed additions and alterations to the UBC 1979 Earthquake Regulations, Chapter 23. These changes are presented for the equipment found in essential facilities only. The proposals are the result of research undertaken during the course of study for this book to assess the vulnerability of essential facilities equipment in the earthquake environment and represent the opinions of this author alone. The proposed additions and alterations have been marked with a solid stripe along the left margin. Sections of the UBC 1979 that are used in their original form are not marked.

Proposed Alterations to the 1979 *Uniform Building Code*

Sec. 2312 (a) **General.**   Every building or structure and every portion thereof shall be designed and constructed to resist stresses produced by lateral forces as provided in this section. Dynamic earthquake tests, mathematical analyses and stress calculations shall be performed as the effect of a force applied horizontally at each floor, roof or equipment attachment point above the base. The effect of a force applied vertically is required of equipment that must remain operational during and after the earthquake.

Structural concepts other than set forth in this section may be approved by the building official when evidence is submitted showing that equipment ductility and energy absorption are provided.

Where prescribed wind loads produce higher stresses, such loads shall be used in lieu of the loads resulting from earthquake forces for the structural system.

(b) **Definitions.**   The following definitions apply only to the provisions of this section:

(Note: Only those definitions added by this author's proposal are included here. All existing UBC 1979 definitions are to remain the same.)

SEISMIC QUALIFICATION PROGRAM FOR BUILDING EQUIPMENT is the overall assessment program to give a reasonable level of assurance that essential facilities will remain operational during and after a significant earthquake.

SEISMIC SPECIFICATION is the detailed plan which describes the requirements of the seismic qualification program.

SEISMIC TEST is the qualification procedure by which building equipment is subjected to a simulated earthquake by a qualified testing laboratory.

(k) **Essential Facilities.**   Essential facilities are those structures or buildings which must be safe and usable for emergency purposes during and after an earthquake in order to preserve the health and safety of the general public. Such facilities shall include but not be limited to:

1.   Hospitals and other medical facilities having surgery or emergency treatment area.
2.   Fire and police stations.
3.   Municipal government disaster operation and communication centers deemed to be vital in emergencies.

Essential facility designs shall include a seismic qualification program that is jointly administered by the design team and owners of the facility. The seismic qualification program shall evaluate all nonstructural components and equipment for earthquake purposes. The evaluation shall include but not be limited to:

1. Communication systems.
2. Emergency power supplies.
3. Fire protection.
4. Laboratory equipment.
5. Life support systems.
6. Occupant circulation systems.
7. Office equipment.
8. Partition walls.
9. Service systems (water, sewage, etc.).
10. Shelved items.
11. Suspended ceilings.

The design team and owner shall determine which equipment is required for the facility operation during and after an earthquake and which is required only on a day-to-day basis. The design team must also obtain design earthquake parameters such as design spectrums from the geotechnical report. The operability of equipment that is required for facility operation and life support or whose failure will directly and adversely affect the function of other required systems or equipment shall be evaluated by either analysis (dynamic or static) or test in the seismic qualification program. Justification for the selected method of qualification (i.e., dynamic test, dynamic analysis, static test, or static analysis) shall be provided as a part of the documentation.

Documentation of all phases of the seismic qualification program must be maintained by the design team for audit by building officials.

Seismic qualification of nonstructural components shall be performed by any of the following means:

1. Design team judgment.
2. Mathematical analysis.
   • Dynamic.
   • Equivalent static coefficient.
3. Prior experience.
4. Seismic simulation.
5. Any combination of the above acceptable to the building officials.

For proper seismic qualification, the design team or owner shall provide the individual, laboratory, or organization performing the seismic qualification with the following:

1. Equipment mounting characteristics.
2. Equipment operational loads.
3. Equivalent static coefficients for equipment requiring anchorage only.
4. Floor response spectra or time history for equipment requiring dynamic solution.

Where the design team cannot provide equivalent static coefficients at equipment locations, for the specified essential facility, the following information may be used. The lateral forces (F_p) on essential facility equipment that is not required for the operation (during and after the earthquake) of the essential facility, or life support, or whose failure will not directly and adversely affect the function of other required systems or equipment shall be:

$$F_p = ZIASHhW_p \hspace{2cm} (1)$$

(replaces UBC 1979
Equation 12-8)

where Z = UBC 1979 zone value

I = importance factor UBC 1979, Table 23-K

A = expected ground acceleration value as defined in Algermissen and Perkins bedrock acceleration map (Figure 3.2b, this book)*

S = soil factor from Table A1.10

H = floor level factor in Table A1.8

h = flexible equipment mounting factor from point of attachment in Table A1.9 for rigid equipment, $h = 1$.

W_p = weight of nonstructural equipment

Flange loads in piping systems are to be reduced by use of flexible connectors at equipment interfaces wherever possible.

Equipment that is required for emergency operation of the facility during and after an earthquake, life support equipment, or equipment whose failure will directly and adversely affect the function of other required systems or equipment shall be shown to be capable of performing the required function before, during, and after the design earthquake. Where operability of the equipment during and after the earthquake is required, a dynamic seismic test is the recommended method of seismic qualification. Seismic analysis methods and anchorage provisions do not necessarily provide assurance of operability.

Where required, dynamic seismic tests shall be conducted in two perpendicular horizontal directions and vertically. Where single axis excitation is employed, the design spectra shall be multiplied by 1.5. Where biaxial tests are employed, the design spectra must be multiplied by 1.2. Equipment shall be operational before, during, and after the seismic excitation. Continuous monitoring of the equipment must be maintained. Equipment shall be mounted on the seismic test machine in a manner that simulates actual installation. The test laboratory shall provide the owner with a detailed report of the seismic test documenting all phases of the test program.

Vibration isolation shall be provided for all resiliently mounted equipment.

Discussion of Proposed Changes to the *Uniform Building Code*

Tables A1.8 and A1.9 and Equation 1 are presented here as a starting point for those designing essential facilities on a limited budget only. The static coefficient method is less exact and where equipment operability must be evaluated, is inappropriate. It is, however, a legitimate approach for evaluating structural integrity and anchorage adequacy. It can only be used for equipment that is not required for the operation of an essential facility during and after an earthquake or for life support.

*While it is recognized that bedrock accelerations do not typify ground accelerations, these values can be used with the modifying soils factors. If expected ground accelerations are known, they may be substituted for "A" and "S" in this equation.

TABLE A1.8. Floor Level Factor

Floor Level	Factor (H)
Basement	1.0
1	1.0
2	1.25
3	1.50
4	1.75
5	2.0
6	2.25
7 or greater	

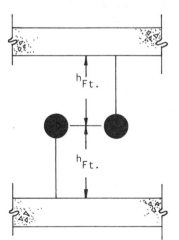

TABLE A1.9. Flexible Equipment Mounting Factor

Equipment Height to Center of Gravity from Anchorage (feet)[a]	Factor (h)
1 or less	1.0
2	1.25
3	1.50
4	1.75
5	2.0
6	2.25
7 or more	2.50

[a] For equipment anchored at base and at upper levels, $h = 1$.

TABLE A1.10. Soil Factor Coefficient

Description	S
Facility located directly on bedrock	1.0
Facility located on well consolidated soil	1.1
Facility located on deep unconsolidated soil	1.5
Facility located on thin veneer of unconsolidated soil	2.0

Tables A1.8 and A1.9 contain equipment acceleration modifiers that increase linearly from their relative bases. The variables are based on observations made from strong motion recordings and damage to equipment in past earthquakes. There is no current evidence that these values are anything more than a rough cut at trying to establish a handle on the very special problems of equipment installation. A considerable amount of work needs to be performed to further refine the modifying values presented here. It is recommended that future research be conducted, preferably seismic testing, to derive modifying values for the equivalent static coefficient approach that can be experimentally substantiated. The programs should include knowledge already obtained in existing seismic qualification programs for the nuclear power industry. The research programs should evaluate a multitude of equipment types, installation practices (rigid/flexible), operational loading factors, and so on. Careful consideration should be given to the various types of structures and earthquakes (duration of strong motion, soil–structure interaction, structural damping, etc.). The results of these tests will reduce the requirements for some equipment installations (e.g., when the equipment natural period falls outside the expected floor response). In other cases the requirements will need to be strengthened (e.g., where the equipment natural period is coincident with the expected floor response).

For an example application of the proposed revisions to UBC 1979, consider the same example used in the general discussion of UBC 1979. The given information is:

- Site—Los Angeles, California.
- Equipment item—cold water pump.
- Weight—500 pounds.
- Location—hospital basement.
- Anchorage—four $\frac{3}{4}$-inch diameter bolt holes in base plate 18 inch on center.

The UBC 1979 proposed approach is:

$F_p = ZIASHhW_p$
$Z = 1$ (zone 4 for Los Angeles, California)
$I = 1.5$ (essential facility)
$A = 0.40$ (40% of gravity horizontal bedrock acceleration, taken from Figure 3.2b, this book)
$S = 1.5$ (deep unconsolidated subsurface geology; obtained from preliminary geologic report)
$H = 1.0$ (equipment located in basement)
$h = 1.25$ (equipment center of gravity is approximately 2 feet above base)
$W_p = 500$ pounds
$F_p = (1)(1.5)(0.40)(1.25)(1.0)(1.5)(500)$
$F_p = 618$ pounds

A comparison of this value with the UBC 1979 value shows that it is approximately 2.5 times that required value.

The real value of the proposed changes to the UBC does not lie with provisions for equipment that only needs to remain anchored. It lies with the provisions provided for equipment that truly must remain operational. The proposals give the UBC the mechanism whereby the "operational requirement" of the existing UBC can be implemented.

SEISMOLOGY COMMITTEE STRUCTURAL ENGINEERS ASSOCIATION OF CALIFORNIA (SEAOC), *RECOMMENDED LATERAL FORCE REQUIREMENTS AND COMMENTARY 1975*

The SEAOC recommendations for lateral force requirements is the "parent code" basis for the UBC 1979 and many others. The UBC 1976 followed the SEAOC recommendations virtually section for section. The UBC 1979 has, however, deviated in some respects. Table 23-J of the UBC 1976 was developed from Table 1-B of the SEAOC recommendations. The UBC 1976 version is more detailed and comprehensive than the SEAOC recommendations. Footnote 4 of SEAOC Table 1-B contains the same disclaimer found in UBC 1979 Table 23-J, footnote 3 (Table A1.6, footnote *d* in this book).

APPLIED TECHNOLOGY COUNCIL ATC 3-06, *TENTATIVE PROVISIONS FOR THE DEVELOPMENT OF SEISMIC REGULATIONS FOR BUILDINGS*

ATC Publication 3-06
NBS Special Publication 510
NSF Publication 78-8

The Applied Technology Council has recently published a proposed seismic code (referred to here as ATC 3-06) designed to be administered at the federal, state, and local levels. ATC 3-06 incorporates both structural and nonstructural provisions, as does UBC 1979. The ATC 3-06 recommendations were distributed in mid-1978 for review and comment. There is, however, no current time schedule set for official adoption of ATC 3-06 at any governmental level. Recognizing the tentative nature of ATC 3-06, the prudent designer may wish to incorporate many of its suggestions to improve the margin of safety of his or her earthquake resistant design. ATC 3-06 contains the most comprehensive provisions for seismic design considerations of building equipment of any of the codes reviewed for this book.

Excerpts

Sec. 1.4.   Seismic Performance

Seismic Performance is a measure of the degree of protection provided for the public and building occupants against the potential hazards resulting from the effects of earthquake motions on buildings. The Seismicity Index and the Seismic Hazard Exposure Group are used in assigning buildings to Seismic Performance Categories. Seismicity Index 4 is associated with the most severe ground shaking expected; Seismic Hazard Exposure Group III is associated with the uses requiring the highest level of protection; Seismic Performance Category D is assigned to provide the highest level of design performance criteria.

1.4.1.   Seismicity Index and Design Ground Motions

The design ground motions are defined in terms of Effective Peak Acceleration or Effective Peak Velocity-Related Acceleration, represented by coefficients A_a and A_v, respectively. The Seismicity Index is related to the Effective Peak Velocity-Related Acceleration Coefficient. The coefficients A_a and A_v and the Seismicity Index to be used in the application of these provisions shall be determined in accordance with the following procedure:

1. Determine the appropriate map area for the building site from Figures C1-3 and C1-4 (Figures A1.4 and A1.5 in this book) for A_a and Figures C1-5 and C1-6 (Figures A1.6 and A1.7 in this book) for A_v.

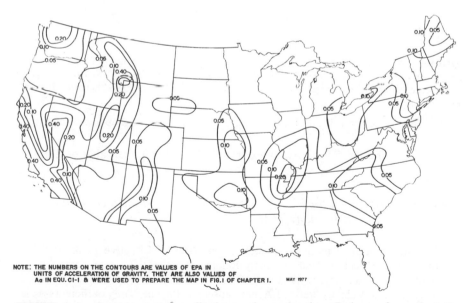

NOTE: THE NUMBERS ON THE CONTOURS ARE VALUES OF EPA IN
UNITS OF ACCELERATION OF GRAVITY. THEY ARE ALSO VALUES OF
Aa IN EQU. CI-I 8 WERE USED TO PREPARE THE MAP IN FIG.I OF CHAPTER I.   MAY 1977

FIGURE A.4.   Contour map for effective peak acceleration. Redrawn from: *Tentative Provisions for the Development of Seismic Regulations for Buildings,* U.S. Department of Commerce, Washington, D.C., 1976, ATZ Publication—ATC3-06, NBS Publication 510, NSF Publication 78-8.

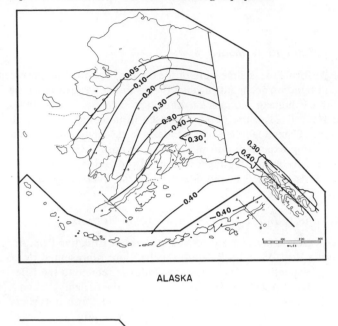

ALASKA

0.20

PUERTO RICO

HAWAII

FIGURE A.5. Contour map for effective peak acceleration. Redrawn from: *Tentative Provisions for the Development of Seismic Regulations for Buildings,* U.S. Department of Commerce, Washington, D.C., 1976, ATZ Publication—ATC3-06, NBS Publication 510, NSF Publication 78-8.

2. Determine the value of A_a and A_v from Table 1-B for the map area found in Step 1.

3. Determine the Seismicity Index from Table 1-B for the value of A_v as determined above.

Alternate Section 1.4.1 for Jurisdictions Which Have Made a Determination of A_a, A_v, and the Seismicity Index:

"The design ground motions are defined in terms of Effective Peak Acceleration and Effective Peak Velocity-Related Acceleration, represented by coefficients A_a and A_v, respectively. The Seismicity Index is related to the effective Peak Velocity-Related Acceleration Coefficient. The coefficients A_a and A_v and the Seismicity Index to be used in the application of these provisions are established as:

$$A_a = \underline{\quad}; A_v = \underline{\quad}; \text{Seismicity Index is } \underline{\quad}."$$

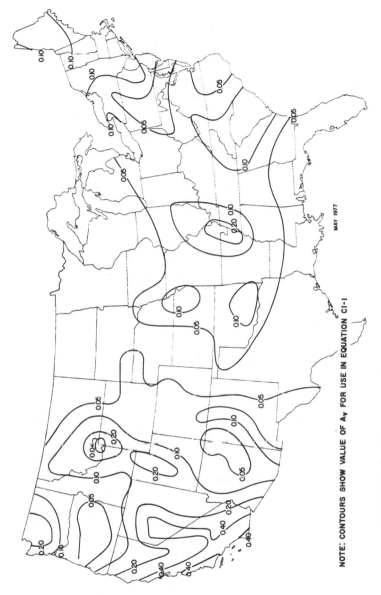

NOTE: CONTOURS SHOW VALUE OF A_v FOR USE IN EQUATION C1-1

FIGURE A.6. Contour map for effective peak velocity-related acceleration coefficient. Redrawn from: *Tentative Provisions for the Development of Seismic Regulations for Buildings*, U.S. Department of Commerce, Washington, D.C., 1976, ATZ Publication—ATC3-06, NBS Publication 510, NSF Publication 78-8.

377

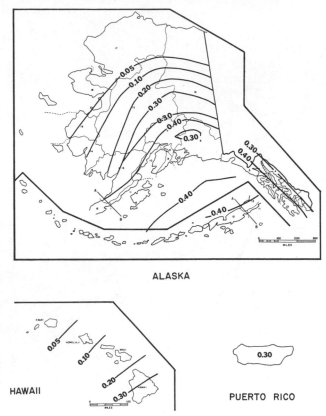

ALASKA

HAWAII

PUERTO RICO

FIGURE A.7. Contour map for effective peak velocity-related acceleration coefficient. Redrawn from: *Tentative Provisions for the Development of Seismic Regulations for Buildings*, U.S. Department of Commerce, Washington, D.C., 1976, ATZ Publication—ATC3-06, NBS Publication 510, NSF Publication 78-8.

1.4.2. Seismic Hazard Exposure Groups

All buildings shall be assigned to one of the following Seismic Hazard Exposure Groups for the purpose of these provisions:

(A) **GROUP III.** Seismic Hazard Exposure Group III shall be buildings having essential facilities which are necessary for post-earthquake recovery. Essential facilities, and designated systems contained therein, shall have the capacity to function during and immediately after an earthquake. Essential facilities are those which have been so designated by the Cognizant Jurisdiction. Access to essential facilities shall conform to the requirements of Sec. 1.4.2(E).

Examples of Possible Group III Facilities:

Fire suppression facilities

Police facilities

Structures housing medical facilities having surgery and emergency treatment areas

Emergency preparedness centers

Power stations or other utilities required as emergency back-up facilities

(B) GROUP II.   Seismic Hazard Exposure Group II shall be buildings having a large number of occupants or buildings in which the occupants' movements are restricted or their mobility is impaired.

Examples of Possible Group II Facilities:

Public assembly for 100 or more persons

Open-air stands for 2,000 or more persons

Day care centers

Schools

Colleges

Retail stores with 5,000 sq ft floor area per floor or more than 35 feet in height

Shopping centers with covered malls, over 30,000 sq ft gross area excluding parking

Offices over 4 stories in height or more than 10,000 sq ft per floor

Hotels over 4 stories in height

Apartment houses over 4 stories in height

Emergency vehicle garages

Ambulatory health facilities

Hospital facilities other than those in Group III

Wholesale stores over 4 stories in height

Factories over 4 stories in height

Printing plants over 4 stories in height

Hazardous occupancies consisting of flammable or toxic liquids including storage facilities for same

(C) GROUP I.   Seismic Hazard Exposure Group I shall be all other buildings not classified in Group III or II.

(D) MULTIPLE USE.   Buildings which have multiple uses shall be assigned the classification of the highest Seismic Hazard Exposure Group which occupies 15 percent or more of the total building area.

(E) PROTECTED ACCESS.   Buildings assigned to Seismic Hazard Exposure Group III shall be accessible during and after an earthquake. Where access is through another structure that structure shall conform to the requirements for Group III. Where access is within 10 feet of side property lines, protection against potential falling hazards from the adjacent property shall be provided.

1.4.3.   Seismic Performance Categories

For the purposes of these provisions all buildings shall be assigned, based on the Seismicity Index established and the Seismic Hazard Exposure Group designated, to a Seismic Performance Category in accordance with Table 1-A.

Any method of analysis or type of construction required for a higher Seismic Performance Category may be used for a lower Seismic Performance Category.

1.4.4. Site Limitation for Seismic Design Performance Category D

No new building or existing building which is, because of change in use, assigned to Category D shall be sited where there is the potential for an active fault to cause rupture of the ground surface at the building.

. . .

1.6.3. Special Testing

. . .

(E) MECHANICAL AND ELECTRICAL EQUIPMENT. For Designated Seismic Systems or components requiring S or G performance ratings in Chapter 8, each component manufacturer shall test or analyze the component and its mounting system or anchorage as required in Chapter 8. He shall submit a certificate of compliance for review and acceptance by the person responsible for the design of the Designated Seismic System and for approval by the Regulatory Agency. The basis of certification required in Sec. 8.3.4 shall be actual test on a shaking table, by three-dimensional shock tests, or by an analytical method using dynamic characteristics and the forces from Formula 8-2, or by more rigorous analysis providing for equivalent safety. The Special Inspector shall examine the Designated Seismic System component and shall determine whether its anchorages and label conform with the certificate of compliance.

SEC. 8.   ARCHITECTURAL, MECHANICAL AND ELECTRICAL COMPONENTS AND SYSTEMS

Sec. 8.1.   General Requirements

The requirements of this Chapter establish minimum design levels for architectural, mechanical, and electrical systems and components recognizing occupancy use, occupant load, need for operational continuity, and the interrelation of structural and architectural, mechanical and electrical components. All architectural, mechanical, and electrical systems and components in buildings and portions thereof shall be designed and constructed to resist seismic forces determined in accordance with this Chapter.
EXCEPTIONS:

1.   Those systems or components designated in Table 8-B or 8-C (Tables A1.12 and A1.13 in this book) for L performance level which are in buildings assigned to Seismic Hazard Exposure Group I and are located in areas with a Seismicity Index of 1 or 2 or which are in buildings assigned to Seismic Hazard Exposure Group II and are located in areas with a Seismicity Index of 1 are not subject to the provisions of this Chapter.

2.   Where alterations or repairs are made the forces on systems or components in existing buildings may be modified in accordance with the provisions of Sec. 13.3.

Seismic Hazard Exposure Groups are determined in Sec. 1.4. Mixed Occupancy requirements are provided in that section.

The seismic force on any component shall be applied at the center of gravity of the component and shall be assumed to act in any horizontal direction. For vertical forces on mechanical and electrical components, see Table 8-C, footnote 2 (Table A1.13, footnote c in this book).

8.1.1. Interrelationship of Components

The interrelationship of systems or components and their effect on each other shall be considered so that the failure of an architectural, mechanical, or electrical system or component of one performance level shall not cause an architectural, mechanical, or electrical system or component of higher level to fail in its performance requirements.

The effect of the response of the structural system and deformational compatibility of architectural, electrical, and mechanical systems or components shall be considered where there is interaction of these systems or components with the structural system.

8.1.2. Connections and Attachments

Architectural, electrical, and mechanical systems and components required to be designed to resist seismic forces shall be attached so that the forces are ultimately transferred to the structure of the building. The attachment shall be designed to resist the prescribed forces.

Friction due to gravity shall not be considered in evaluating the required resistance to seismic forces.

The design documents shall include sufficient information relating to the attachments to verify compliance with the requirements of this Chapter.

8.1.3. Performance Criteria

The performance criteria for architectural, mechanical, and electrical components and systems are listed in Table 8 A (Table A1.11 in this book) for use in Formulas 8-1 and 8-2 and Tables 8-B and 8-C.

These components and systems shall be designed to meet the performance characteristic levels established in Tables 8-B and 8-C.

TABLE A1.11.   Performance Criteria (Table 8-A)

Designation[a]	Performance Characteristic Level	P
S	Superior	1.5
G	Good	1.0
L	Low	0.5

[a]See Tables 8-B and 8-C [Tables A1.12 and A1.13 in this book].

TABLE A1.12. Seismic Coefficient (C_c) and Performance Characteristic Levels Required for Architectural Systems or Components (Table 8-B)[a]

Architectural Components	C_c Factor	Required Performance Characteristic Levels Seismic Hazard Exposure Group		
		III	II	I
Appendages				
Exterior nonbearing walls	0.9	S	G[c]	L[d]
Wall attachments	3.0	S	G[c]	L[d]
Veneers	3.0	G	G[b]	L
Roofing units	0.6	G	G[c]	NR
Containers and miscellaneous components (freestanding)	1.5	G	G	NR
Partitions				
Stairs and shafts	1.5	S	G[d]	G
Elevators and shafts	1.5	S	L[d]	L[f]
Vertical shafts	0.9	S	L[d]	L[g]
Horizontal exits including ceilings	0.9	S	S	G
Public corridors	0.9	S	G	L
Private corridors	0.6	S	L	NR
Full-height area separation partitions	0.9	S	G	G
Other Pull-height partitions	0.6	S	L	L
Partial-height partitions	0.6	G	L	NR
Structural fireproofing	0.9	S	G[d]	L[g]
Ceilings				
Fire-rated membrane	0.9	S	G[d]	G
Nonfire-rated membrane	0.6	G	G	L
Architectural equipment—ceiling, Wall, or floor-mounted	0.9	S	G	L

[a] See Table A1.11 for S, G, and L designations. NR = not required.

[b] May be reduced one performance level if the area facing the exterior wall is nominally inaccessible for a distance of 10 feet plus 1 foot for each floor of height.

[c] May be reduced one performance level if the area facing the exterior wall is nominally inaccessible for a distance of 10 feet and building is only one story.

[d] Shall be raised one performance level if building is more than four stories or 40 feet in height.

[e] Shall be raised one performance level if building is in an urban area.

[f] May be reduced to NR if building is less than 40 feet in height.

[g] Shall be raised one performance level for an occupancy containing flammable gases, liquids, or dust.

TABLE A1.13. Seismic Coefficient (C_c) and Performance Characteristic Levels Required for Mechanical/Electrical Components (Table 8-C)[a]

Mechanical/Electrical Components[b]	C_c Factor[c]	Required Performance Characteristic Levels — Seismic Hazard Exposure Group III	II	I
Emergency electrical systems (code required) Fire and smoke detection system (code required) Fire supperssion systems (code required) Life safety system components	2.00	S	S	S
Boilers, furnaces, incinerators, water heaters, and other equipment using combustible energy sources or high temperature energy sources, chimneys, flues, smokestacks, and vents				
Communication systems Electrical bus ducts and primary cable systems Electrical motor control centers, motor control devices, switchgear, transformers, and unit substations Reciprocating or rotating equipment Tanks, heat exchangers, and pressure vessels Utility and service interfaces	2.00	S	G	L
Machinery (manufacturing and process)	0.67	S	G	L
Lighting fixtures	0.67[d]	S	G	L
Ducts and piping distribution systems				
Resiliently supported	2.00	S	G	NR
Rigidly supported	0.67[e]	S	G	NR
Electrical panelboards and dimmers	0.67	S	G	NR
Conveyor systems (nonpersonnel)	0.67	S	NR	NR

[a]See Table A1.11 for S, G, and L designations. NR = not required.

[b]Where mechanical or electrical components are not specifically listed in Table A1.13 the designer shall select a similarly listed component, subject to the approval of the authority having jurisdiction, and shall base the design on the performance and C_c values for the similar component.

[c]C_c values listed are for horizontal forces. C_c values for vertical forces shall be taken as one-third of the horizontal values.

[d]Hanging or swinging fixtures shall use a C_c value of 1.5 and shall have a safety cable attached to the structure and the fixture at each support point capable of supporting four times the vertical load.

[e]Seismic restraints may be omitted from the following installations:

a. Gas piping less than 1-inch inside diameter.

b. Piping in boiler and mechanical rooms less than $1\frac{1}{4}$ inches inside diameter.

c. All other piping less than $2\frac{1}{2}$ inches inside diameter.

d. All electrical conduit less than $2\frac{1}{2}$ inches inside diameter.

e. All rectangular air-handling ducts less than 6 square feet in cross-sectional area.

f. All round air-handling ducts less than 28 inches in diameter.

g. All piping suspended by individual hangers 12 inches or less in length from the top of the pipe to the bottom of the support for the hanger.

h. All ducts suspended by hangers 12 inches or less in length from the top of the duct to the bottom of the support for the hanger.

Sec. 8.2.   Architectural Design Requirements

8.2.1.   General

Systems or components listed in Table 8-B and their attachments shall be designed and detailed in accordance with the requirements of this Chapter. The designs or criteria for systems or components shall be included as part of the design documents.

8.2.2.   Forces

Architectural systems and components and their attachments shall be designed to resist seismic forces determined in accordance with the following formula:

$$F_p = A_v C_c P W_c \qquad (8\text{-}1)$$

where F_p = The seismic force applied to a component of a building or equipment at its center of gravity.

C_c = The seismic coefficient for components of architectural systems as given in Table 8-B (dimensionless).

W_c = The weight of a component of a building or equipment.

A_v = The seismic coefficient representing the Effective Peak Velocity-Related Acceleration as determined in Sec. 1.4.

P = Performance criteria factor as given in Table 8-A (dimensionless).

EXCEPTIONS:
When positive and negative wind loads exceed F_p for nonbearing exterior walls, these loads shall govern the design. Similarly, when the code horizontal loads exceed F_p for interior partitions, these loads shall govern the design.

8.2.3.   Exterior Wall Panel Attachment

Attachment of exterior wall panels to the building seismic resisting system shall have sufficient ductility and provide rotational capacity needed to accommodate the design story drift determined in accordance with Sec. 4.6.1.

8.2.4.   Component Deformation

Provisions shall be made in the architectural system or component for the design story drift Δ as determined in Sec. 4.6.1. Provision shall be made for vertical deflection due to joint rotation of cantilever members.
EXCEPTION:
Components assigned an L performance factor in Table 8-B may provide for a design story drift of $\Delta/2$.

8.2.5.   Out-of-Plane Bending

Transverse or out-of-plane bending or deformation of a component or system composed of basically brittle materials which are subject to forces as determined in Formula 8-1 shall not exceed the deflection capability of the material.

Sec. 8.3. *Mechanical and Electrical Design Requirements*

8.3.1. *General*

Systems or components listed in Table 8-C and their attachments shall be designed and detailed in accordance with the requirements of this Chapter. The designs or criteria for systems or components shall be included as part of the design documents.

An analysis of a component supporting mechanism based on established principles of structural dynamics may be performed to justify reducing the forces determined in Sec. 8.3.2.

Combined states of stress, such as tension and shear in anchor bolts, shall be investigated in accordance with established principles of mechanics.

8.3.2. *Forces*

Mechanical and electrical components and their attachments shall be designed for seismic forces determined in accordance with the following formula:

$$F_p = A_v C_c P a_c a_x W_c \tag{8-2}$$

where F_p, A_v, P, and W_c are as defined previously in Sec. 8.2.2.

$\quad C_c$ = The seismic coefficient for components of mechanical or electrical systems as given in Table 8-C (dimensionless).

$\quad a_c$ = The amplification factor related to the response of a system or component as affected by the type of attachment, determined in Sec. 8.3.2(A).

$\quad a_x$ = The amplification factor at level x related to the variation of the response in height of the building.

The amplification factor, a_x, shall be determined in accordance with the following formula:

$$a_x = 1.0 + (h_x/h_n) \tag{8-3}$$

where h_x = The height above the base to level x

$\qquad h_n$ = The height above the base to level n

(A) ATTACHMENT AMPLIFICATION. The attachment amplification factor, a_c, shall be determined as follows:

For fixed or direct attachment to buildings: $a_c = 1$.

For resilient mounting system:

$\quad$ with seismic activated restraining device $a_c = 1$.

$\quad$ with elastic restraining device:

$\qquad$ if $T_c/T < 0.6$ or $T_c/T > 1.4$ $\quad\quad a_c = 1$.

$\qquad$ if $T_c/T \geqslant 0.6$ or $\leqslant 1.4$ $\quad\quad a_c = 2$ minimum.*

$\qquad$ if mounted on the ground or on a slab in direct contact with the ground $\quad a_c = 2$.

*See Sec. 8.3 of Commentary.

The value of the fundamental period, T, shall be the value used in the design of the building as determined in accordance with Sec. 4.2 or Sec. 5.4.

The fundamental period of the component and its attachment, T_c, shall be determined in accordance with the following formula:

$$T_c = 0.32 \sqrt{\frac{W_c}{K}} \qquad (8\text{-}4)$$

where K = The stiffness of the equipment support attachment determined in terms of load per unit deflection of the center of gravity (lbs./in.) as follows:

For stable resilient attachments, K = spring constant.

For other resilient attachments, K = slope of the load/deflection curve at the point of loading.

In lieu of Formula 8-4, properly substantiated values for T_c derived using experimental data or any generally accepted analytical procedure may be used.

8.3.3. Attachment Design

Fixed or direct attachments shall be designed for the forces determined in Sec. 8.3.2 and in conformance with Chapters 9, 10, 11, or 12 for the materials comprising the attachment.

Resilient mounting devices shall be of the stable type. Restraining devices shall be provided to limit the horizontal and vertical motion, to inhibit the forces from forcing the resilient mounting system into resonance, and to prevent overturning. Elastic restraining devices shall be designed based upon the forces obtained from Formula 8-2 or in accordance with the dynamic properties of the component and the structure to which it is attached. Horizontal and vertical elastic restraining devices shall be designed to decelerate the component or system on contact at a rate which will not generate forces in excess of those calculated from Formula 8-2.

8.3.4. Component Design

When the direct attachment method is to be used for components with performance characteristic levels of S or G in areas with Seismicity Index 3 or 4, the designer shall require certification from the manufacturer that the components will not sustain damage if subjected to forces equivalent to those resulting from Formula 8-2.

When resilient mounting systems are used for components with performance criteria levels S or G both the mounting systems and the components shall require the certification stated above. Such systems shall be of the stable type.

Testing and certification shall be in accordance with the requirements of Sec. 1.6.3.

8.3.5. Utility and Service Interfaces

The utility or service interface of all gas, high-temperature energy and electrical supply to buildings housing Seismic Hazard Exposure Groups II and III and located in areas having a Seismicity Index of 3 or 4 shall be provided with shutoff devices located at the building side of the interface. Such shutoff devices shall be activated either by a failure within a system being supplied or by a mechanism which will operate when the ground motion exceeds $0.5\,A_a$ times the acceleration of gravity.

General Discussion of ATC 3-06

As is stated in the introductory text of Appendix 1, it is this author's opinion that ATC 3-06 contains the most comprehensive and workable provisions for seismic design considerations of building equipment that has been presented to date. Its attributes are many, and its failings appear to this author to be few.

ATC 3-06 is the first major published architectural specification to consider the interrelationship of components. The failure of one component must be considered so as to not adversely affect the operability of another component that is in a higher seismic performance category. An example of such an occurrence might be a large unrestrained wheeled centrifuge (suggested seismic category "D") that smashes into and seriously damages a blood bank refrigerator (suggested seismic category "A") in a hematology laboratory.

The seismic hazard exposure group (Section 1.4.2 ATC 3-06) is considerably more comprehensive than the equivalent UBC 1979 importance factor approach. The performance criteria established by ATC 3-06 lead the design team into the systems approach to qualification and recognize that some equipment is relatively more important to the operation of a given facility type than other equipment items.

The forces applied to architectural, electrical, and mechanical equipment (ATC 3-06 Equations 8-1 and 8-2) yield values that correspond well with the values obtained from the proposed changes to the UBC 1979. This author arrived at the proposed changes to the UBC independently of ATC 3-06 during late 1977 and early 1978, approximately the same time frame as ATC 3-06, which was distributed in mid-1978. Vibration isolated equipment is not, however, adequately addressed by ATC 3-06; a serious flaw.

ATC 3-06 is the first general use architectural seismic code or specification to address seismic testing of equipment. ATC 3-06 states in Section 1.6.3(E)

> For Designated Seismic Systems or components requiring S or G performance ratings in Chapter 8, each component manufacturer shall test or analyze the component and its mounting system or anchorage as required in Chapter 8.

While it is generally good that ATC 3-06 addresses the seismic test for building equipment, it fails to recognize the real use of the seismic test. Requiring manufacturers to test their equipment or component for base anchorage is generally a misuse of the seismic test. Testing procedures are generally expensive and should be reserved for proving the operability of critical equipment, not proving that it will remain anchored. Anchorage can usually be best and most economically determined by analysis, not seismic testing.

For an example application of ATC 3-06 equipment qualification analysis, consider the same example used in the General Discussion of UBC 1979. The given information is:

- Site—Los Angeles, California.
- Equipment item—cold water pump.
- Weight—500 pounds.
- Location—hospital basement.
- Anchorage—4 $\frac{3}{4}$-inch diameter bolt holes in base plate 18 inch on center.

The ATC 3-06 equivalent static coefficient analysis formula is:

$$F_p = A_v C_c Pa_c a_x W_c \qquad (8\text{-}2)$$

where A_v = 0.4 (from Figure C1-5)

$\quad C_c$ = 2.0 (from Table 8-C, reciprocating or rotating equipment)

$\quad P$ = 1.5 (from Tables 8-A and 8-C, seismic hazard exposure group III, superior performance required)

$\quad a_c$ = 1 (from Section 8.3.2A)

$\quad a_c$ = 1(a_x = 1 + $[h_x/h_n]h_x$ = 0 basement h_h = 2 feet to center of gravity

$\quad W_c$ = 500 pounds

$\quad F_p$ = (0.4)(2.0)(1.5)(1)(1)(500)

$\quad F_p$ = 600 pounds

A comparison of this value with those obtained in the UBC 1979 examinations shows that ATC 3-06 has lateral load requirements approximately 2.5 times the UBC 1979 value obtained. Comparison of the value obtained from this author's proposed changes to UBC 1979 and the ATC 3-06 value shows that they are very close. The UBC 1979 proposed change value would increase markedly if the pump were located say on the fourth floor of the hospital, as would the ATC 3-06 value, and the base anchorage values would still be closely comparable.

Evaluation*

Section 8.3 for systems or components is flawed due to its failure to achieve a worthy but impossible goal. It misses an opportunity to specify proper clearances and shock cushions in restraining devices and it fails to include formulas for standardized manual determination of loads on anchors and restraints.

*This section, and pages 389 through 392 provided courtesy of Donald Benkert, California Dynamics Corporation.

Impossible Goal

Section 8.3.2 of ATC 3-06 quite properly attempts to provide a standard technique to determine F_p, the design static force considered to be applied at the CG of mechanical and electrical components. Unfortunately, coefficient a_c, "the amplification factor related to the response of a system or component as affected by the type of attachment," is improperly defined. It is obvious that a_c was intended to be an easily determined factor to account for crucial dynamic effects in the design of seismic protection. This is a worthy but impossible goal. Instead of assuming severe effects and adjusting coefficient C_c accordingly, the technique involves a futile attempt to determine if a_c is 1, 2, or 2 minimum* "For resilient mounting systems with elastic restraining devices." The asterisk leaves the reader in limbo with a grossly oversimplified equation on page 422 of ATC 3-06 which is improper in this application. Since validity is destroyed by incompleteness, each of the many simplifications found necessary decreases the resemblance of validity of a_c resulting from use of plate C8-1 on page 435. The period ratios on page 80 are erroneously pursued as if the periods existed in steady state. This is an unrealistic target for an isolated and restrained component since motion in the extreme is a jerky rocking in all directions with scant evidence of the "fundamental period of the component and attachment T_c" determined with formula 8-4. Improper results are guaranteed with formula 8-4 by exclusion of numerous important parameters as well as the crucial clearances and cushion compliances which are paramount in describing the extreme motion of properly protected components. Thus a_c is a failed attempt to achieve an impossible goal. Had the numerous parameters been included, the complexities would have been increased enormously and the dynamic analysis would have become impossible to handle without extremely involved computer techniques.

Recommendations

Obviously, the extreme complexities of a complete dynamic analysis do not serve as license to ignore physical realities. On the other hand, proven techniques, which are available at a negligible cost, could satisfactorily deemphasize the physical realities and substantially reduce the problem from dynamic to static. It's likely that a complete dynamic analysis would do little more than confirm the need for these proven techniques using proper shock cushions and clearances. Since there is little penalty in satisfying a possible need for protection, providing the cushioning is preferable to expensive analysis which would almost certainly confirm the need for shock cushioning protection.

1. Since a_c is a failed attempt at an impossible goal, a_c **should be abandoned and coefficients C_c should be adjusted accordingly considering extreme conditions mitigated as in 2 and 3 below.**

2. Because the scheme is protection from, not avoidance of an earthquake, elements of restraints are expected to collide. Although increasing the clear spaces between elements within the restraints leads to a reduced tendency to collide, it also tends to reduce protection. Widely spaced elements that do not collide, also do not restrain, and in such cases the component is unprotected. Recognizing collision as inevitable, we minimize severity by minimizing acceleration distance. Thus, clearances must be as small as possible without compromising isolation. **Proper clearances should be clearly and unmistakably specified.**

3. Resilient shock cushions are able to absorb work through significant distances and quickly return to their original positions without a failure that minimizes shock loads within the restraining system. Without proper resilient shock cushions for protection, kinetic energy in the moving component may exceed the capacity of the restraining system to absorb work and result in failure. Shock cushions are inexpensive, reliable, and highly beneficial shock mitigators. **Proper shock cushions should be clearly and unmistakably specified.**

4. Section 8.3.1 recognizes the need that "anchor bolts, . . . be investigated" but fails to provide badly needed formulas for standardized manual determination of loads on anchor bolts. **Formulas based upon the well-received techniques shown on the "Anchors" form are recommended to satisfy this need.** The form deals with two anchor bolt configurations. Appropriate adjustments must be made for restraints with more than two anchor bolts.

5. Section 8.3.3 recognizes the need that "restraining devices . . . be designed based upon the forces obtained" but fails to provide badly needed formulas for standardized manual determination of loads on restraining devices. **Formulas based upon the well-received techniques shown on the "Isolator Restraint" form are recommended to satisfy this need.**

Structural Integrity of Restraining Elements for Earthquake Protection
Part 1: Isolator Restraint

Job _____

Equipment Supplier _____ Equipment _____

Isolator Restraints _____ Date _____

Anchor Type _____ Size _____

Shear Rating _____ Pullout Rating _____ Basis _____

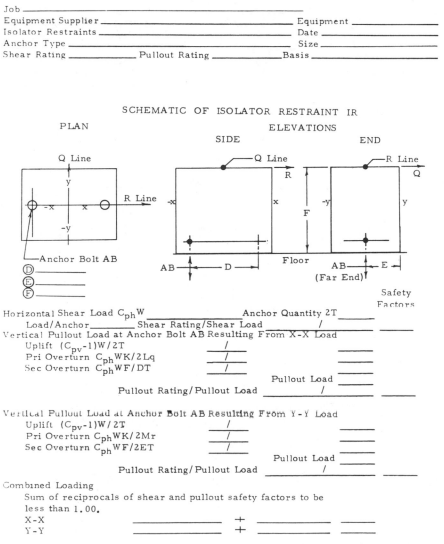

SCHEMATIC OF ISOLATOR RESTRAINT IR

PLAN   ELEVATIONS

SIDE   END

Safety Factors

Horizontal Shear Load $C_{ph}W$ _____ Anchor Quantity 2T _____

    Load/Anchor _____ Shear Rating/Shear Load _____ / _____ _____

Vertical Pullout Load at Anchor Bolt AB Resulting From X-X Load

    Uplift $(C_{pv}-1)W/2T$ _____ / _____ _____

    Pri Overturn $C_{ph}WK/2Lq$ _____ / _____ _____

    Sec Overturn $C_{ph}WF/DT$ _____ / _____

                                  Pullout Load _____

             Pullout Rating/Pullout Load _____ / _____ _____

Vertical Pullout Load at Anchor Bolt AB Resulting From Y-Y Load

    Uplift $(C_{pv}-1)W/2T$ _____ / _____ _____

    Pri Overturn $C_{ph}WK/2Mr$ _____ / _____ _____

    Sec Overturn $C_{ph}WF/2ET$ _____ / _____

                                  Pullout Load _____

             Pullout Rating/Pullout Load _____ / _____ _____

Combined Loading

    Sum of reciprocals of shear and pullout safety factors to be less than 1.00.

    X-X _____ + _____ _____

    Y-Y _____ + _____ _____

Courtesy California Dynamics Corp.

391

Structural Integrity of Restraining Elements for Earthquake Protection
Part 2: Anchors

Job _____

Equipment Supplier _____ Equipment _____

Equipment Weight Ⓦ _____ Date _____

Isolator Restraints _____ Quantity Ⓣ _____

Protection Requirement Horizontal Ⓒₚₕ _____ Vertical Ⓒₚᵥ _____

X-X Rating _____ Y-Y Rating _____ Z-Z Rating _____

Basis _____

SCHEMATIC OF EQUIPMENT

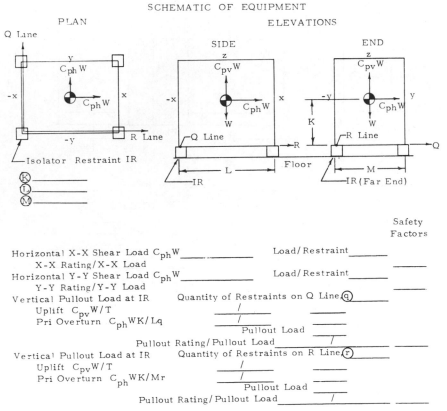

PLAN   ELEVATIONS

Ⓚ _____
Ⓛ _____
Ⓜ _____

Safety
Factors

Horizontal X-X Shear Load $C_{ph}W$ _____   Load/Restraint _____ _____
    X-X Rating/X-X Load

Horizontal Y-Y Shear Load $C_{ph}W$ _____   Load/Restraint _____ _____
    Y-Y Rating/Y-Y Load

Vertical Pullout Load at IR   Quantity of Restraints on Q Line ⓠ _____
    Uplift $C_{pv}W/T$ / _____ _____
    Pri Overturn $C_{ph}WK/Lq$ / _____
                       Pullout Load _____
         Pullout Rating/Pullout Load _____ / _____

Vertical Pullout Load at IR   Quantity of Restraints on R Line ⓡ _____
    Uplift $C_{pv}W/T$ / _____ _____
    Pri Overturn $C_{ph}WK/Mr$ / _____
                       Pullout Load _____
         Pullout Rating/Pullout Load _____ / _____

Courtesy California Dynamics Corp.

392

STATE OF CALIFORNIA, CALIFORNIA ADMINISTRATIVE CODE, TITLE 24, BUILDING STANDARDS, PART 3 (CAC 24-3), "BASIC ELECTRICAL REGULATIONS"

There are no specific requirements in CAC 24-3 that deal directly with seismic provisions. Permanent pendant light fixtures, however, with stems more than 12 inches long are required to have bracing against lateral displacements. This is the only reference in CAC 24-3 to lateral loading. Other requirements call for cable trays, boxes, cabinets, raceways, conduit, and so on, to be secured in place or adequately supported. Emergency generator sets and storage battery provisions are discussed; however, there is no reference to anchorage against lateral loading for these assemblies. Emergency lighting equipment is required to be permanently fixed in place by CAC 24-3, again without any mention of seismic provisions. Emergency lighting is, unfortunately, seldom anchored and is typically left sitting on a shelf or filing cabinet waiting to fall.

STATE OF CALIFORNIA, CALIFORNIA ADMINISTRATIVE CODE, TITLE 24, BUILDING STANDARDS, PART 4 (CAC 24-4), "BASIC MECHANICAL REGULATIONS"

The basic mechanical regulations do not make any specific references to seismic loading. Many of the equipment categories, however, do require anchorage. Space cooling units are required to be bolted down and they must remain at least 6 inches from any adjoining construction. Whether or not this requirement was imposed to restrict hammering of equipment in the seismic environment is not known. It will serve that function for many pieces of equipment whether or not it was specifically intended for the seismic environment. Hoods, above ground tanks, gas lines, and so on, all require anchorage or support as provided by CAC 24-4. Gas (fuel) lines to emergency power supplies are required to be supported according to the line diameter. No mention is made in this code for the required type of connection. Past earthquakes have shown rigid connections at the day tank and generator set to perform unfavorably during an earthquake. Flexible connections, on the other hand, seem to perform quite well.

STATE OF CALIFORNIA, CALIFORNIA ADMINISTRATIVE CODE, TITLE 24, BUILDING STANDARDS, PART 5 (CAC 24-5), "BASIC PLUMBING REGULATIONS"

There are no specific seismic requirements for piping in CAC 24-5. However, all piping (vertical and horizontal) requires supporting and/or anchorage. Attention needs to be placed on piping systems where they cross seis-

mic joints and between floors where story displacements occur. Consideration also needs to be given where plumbing passes through bearing walls,
nonbearing walls, and partition walls. Piping systems critical to the facility
operation may require snubbing systems similar to those found in nuclear
power stations.

STATE OF CALIFORNIA, CALIFORNIA ADMINISTRATIVE CODE, TITLE 24, BUILDING STANDARDS, PART 6 (CAC 24-6), "SPECIAL BUILDING REGULATIONS"

CAC 24-6 is based on the *Uniform Building Code* (prior to 1976). The
nonstructural requirements of the UBC 1979 are much more stringent. It is
expected that CAC 24-6 will adopt the UBC 1979 requirements in the future.

STATE OF CALIFORNIA, CALIFORNIA ADMINISTRATIVE CODE, TITLE 24, BUILDING STANDARDS, PART 7 (CAC 24-7), "ELEVATOR SAFETY REGULATIONS"

Excerpts

The following are excerpts from CAC 24-7. Portions not pertaining to seismic conditions have been omitted.

Article 6.   Definitions

3009.   Definitions.

· · ·

   (20.1) Earthquake Protection Devices. A device or group of devices which
serve to regulate the operation of an elevator in a predetermined manner during
or after an earthquake.
      (A) Seismic Switch. A device actuated by building movement to provide
   information to the control that an earthquake is causing movement of the
   building.
      (B) Collision Switch. A device actuated by the car or counterweight to
   provide information to the control that a collision between the car and the
   counterweight is imminent.
      (C) Derailment Switch. A device actuated by the derailment of the
   counterweight at any point in the hoistway to provide information to the
   control that the counterweight has left its guides.

· · ·

3015.   Machinery and Sheave Beams, Supports, and Foundations.

   (a) Beams and Supports Required.
      (1) *Machines, machinery, and sheaves shall be so supported and main-
   tained in place as to effectually prevent any part from becoming loose or
   displaced under the conditions imposed in service.*

· · ·

(c) Securing of Machinery and Equipment to Beams, Foundations or Floors.
   (1) Machinery or equipment shall be secured to and supported on or from the top of overhead beams or floors.

· · ·

   (3) The fastenings including vibration isolation units and supporting structures used to attach *controllers, motor generator sets, compensating rope sheave assemblies,* machines, machine beams, and sheaves to the building shall conform to Section 3111(c)(3).
      Exception: A period of seven years from the effective date of this order will be allowed for existing buildings to comply with the requirements of Order 3015(c)(3).

· · ·

Article 8. *Machinery and Equipment for Power Cable-driven Passenger and Freight Elevators*

3030. *Car and Counterweight Guide Rails, Rail Supports, and Fastenings.*

· · ·

(k) Guide Rail Brackets and Building Supports.

· · ·

   (C) Withstand seismic forces created by accelerations of 0.5g horizontally acting on the car and/or counterweight in their most adverse position in relation to any bracket without deflecting more than ¼" and without exceeding 88% of the yield strength of the material used.
   (2) Guide rail brackets shall be secured to their supporting structure by means of bolts, rivets, or by welding to withstand forces described in Section 3030(k)(1)(C). Fastening bolts and bolt holes in brackets and their supporting beams shall conform to the requirements of Section 3030(1). Welding shall conform to Section 3030(g).

· · ·

3032. *Counterweights.*

(a) Frames. *Counterweight weight sections shall be mounted in structural or formed metal frames, so designed as to retain the weights securely in place.*

· · ·

(c) Guiding of Counterweights. *All elevator counterweights shall run in guides.*
   Exception: Existing counterweights running in boxes.
   (1) Counterweight frames shall be guided on each guide rail by upper and lower guiding members attached to the frame. *The guiding members or auxiliary guiding members and their attachment to the counterweight frame shall be designed to withstand seismic forces of not less than 0.5g horizontally. The clearances between the machined faces of the rail and auxiliary guiding members shall be not more than $\frac{3}{16}$". The engagement of the rail shall not be less than the dimension of the machined side face of the rail.*
      Exception: A period of seven years from the effective date of this order will be allowed for existing buildings to comply with the requirements of Order 3032(c)(1).

· · ·

3041.   Emergency and Signal Devices.

.   .   .

(d) Earthquake Emergency Operation.

(1) Passenger elevators with automatic operation and counterweights shall be provided with earthquake protective devices of the following types.

Exception: *Elevators whose car and counterweight guiding system including rails, brackets and guiding shoes whose equipment fastenings and attachments to the building structural members have been properly certified to the Division, by an engineer qualified under the Civil and Professional Engineers Act, to be designed and built to withstand the static and dynamic seismic forces for which the building was designed.*

(A) Elevators with drum machines operating at any speed and traction machines operating at rated speeds of more than 150 f.p.m. shall be provided with either a seismic switch device or a derailment switch device.

(B) Elevators with traction machines with rated speeds of not more than 150 f.p.m. shall be provided with a collision switch device.

Exceptions:

1. *Elevators provided with either a seismic switch device or a derailment switch device.*
2. *Elevators with traction machines with counterweights located or restrained to prevent the car and counterweight colliding.*

(C) Elevators with traction machines arranged to operate under emergency conditions after activation of either a seismic switch device or a derailment switch device shall be provided with a collision switch device.

Exceptions:

1. *Elevators with traction machines with counterweights located or restrained to prevent the car and counterweight colliding.*
2. *Elevators equipped with a derailment switch that continuously monitors the position of the counterweight and therefore acts as a collision switch.*

(2) Passenger elevators with traction machines, counterweights, and selective collective or group automatic operation shall upon activation of an earthquake protective device, and if in motion, either . . .

(A) Slow to a speed not greater than 150 f.p.m. and proceed to the next floor in the direction of the travel and stop.

Exception: *Elevators operating in a hoistway with more than 36 feet between landings shall not proceed to the next floor in the direction of travel if the car must pass the counterweight.*

(B) Stop and then proceed to the next floor at a speed not greater than 150 f.p.m. in a direction away from the counterweight.

(3) Passenger elevators with traction machines having automatic operation other than selective collective or group automatic shall, if in motion, upon activation of an earthquake protection device, stop.

Exception: *Elevators with traction machines that comply with 3041(d)(2).*

(4) Passenger elevators with counterweights and drum machines shall, if in motion, upon activation of an earthquake protective device stop.

(5) Freight elevators and passenger elevators with a designated operator in the car, shall have an audible signal in the car to indicate that a seismic switch device or a derailment switch device has been activated.

Exceptions:

1. *Freight elevators with rated speeds of not more than 150 f.p.m. whose counterweight rails comply with Section 3030(f)(3)(A) and 3032(c)(1).*
2. *Elevators arranged to comply with Section 3041(d)(2) or Section 3041(d)(3) or Section 3041(d)(4).*

(6) Cars stopped by an earthquake protection device shall be operable at not more than 150 f.p.m. from the car top operating station as described in Section 3040(a)(4) if so equipped.

(7) Activation of a seismic switch device or momentary activation of a derailment switch device shall prevent operation of the car by the emergency service key described in 3041(c) or a hospital emergency service key at a speed greater than 150 f.p.m.

(8) Activation of a collision switch device or continuous activation of a derailment switch device shall prevent operation of the car except from the car top operating station.

   Exception: Cars stopped by activation of the collision switch or derailment switch may be operable from the emergency service switch described in Section 3041(c) in the direction away from the counterweight.

(9) A collision switch shall, upon activation, stop an elevator traveling at a speed of 150 f.p.m. before the car meets the counterweight.

(10) Elevators not in operation shall remain at the landing. Elevators shall upon reaching a landing remain at the landing unless operated by the emergency service key described in Section 3041(c).

(11) Cars with power operated doors shall upon reaching a landing cause their doors to open and remain open unless operated by the emergency service key described in Section 3041(c).

(12) An earthquake sensing device shall activate upon excitation in a horizontal or vertical direction of not more than 0.15g.

(13) An identified momentary reset button or switch for each car located in the control panel in the machine room shall be provided for elevators equipped with a seismic switch or a derailment switch.

(14) Cars stopped by an earthquake protection device shall remain stopped in the event of a power failure and subsequent restoration of power. The functions performed by the electrical protective devices required by Section 3040(b) shall not be cancelled by the earthquake protection device.

(15) Earthquake protection devices with the exposed live parts in the hoistway shall operate at not more than 24 volts Root Mean Square A.C. or 24 volts D.C. above or below ground potential and shall not be capable of supplying more than $\frac{1}{2}$ ampere when short circuited.

(16) Earthquake protection devices shall be of a fail safe type or shall include a dual system arranged to prevent energizing the sensing portion unless the complete system is intact.

(17) Earthquake protection devices shall be arranged to be checked for satisfactory operation and shall be calibrated at intervals recommended by the manufacturer.

   Exception: A period of seven years from the effective date of this order will be allowed for existing buildings to comply with the requirements of Order 3041(d).

History: **1.** New subsection (d) filed 3-30-76; effective thirtieth day thereafter (Register 76, No. 14). Approved by State Building Standards Commission 3-26-76.

        2. Repealed and new subsection (c) filed 7-30-76; effective thirtieth day thereafter (Register 76, No. 31). Approved by State Building Standards Commission 7-16-76.

. . .

3053. Machine and Sheave Beams, Supports, and Foundations and Pits.

. . .

(c) The pump unit tank shall be supported and maintained in place to conform to Section 3111(c)(3).

. . .

3111. Stresses and Deflection in Machinery and Sheave Beams and Their Supports.

. . .

    (3) The fastenings used to attach equipment, except rail brackets, to the supporting structure shall be designed to withstand seismic forces of $1.0g$ horizontally and $0.5g$ vertically acting simultaneously when such fastenings are rigid or when fastenings use rubber or similar material for vibration isolation of equipment. Fastenings using springs for vibration isolation of equipment shall be designed to withstand forces double those for rigid fastenings.

    The stresses in parts or structural members made of steel shall not exceed 88% of the yield strength of the material used in the fastenings.

General Discussion of CAC 24-7

CAC 24-7 has many direct and indirect seismic requirements that relate to elevators and their machinery. Past earthquake experience has dictated the upgrading of CAC 24-7. Machinery is now required to be supported and/or anchored to its foundation. Elevator equipment must also be equipped with seismic switches that control the function of an elevator after an earthquake. CAC 24-7 allows building owners 7 years from the date of adoption of the earthquake provisions to comply with the regulations. This means that elevator owners must have "retrofit" contracts under way by October 1982.

    Montgomery Elevator estimates that there are approximately 12,000 elevators in California that are subject to the seismic retrofit program. If it is assumed that the average price for retrofit is $2500 per elevator, the estimated value of work required is $30 million.

    CAC 24-7 directs most of the attention toward cable type elevators with traction machines rather than toward hydraulic type elevators. The new requirements impose horizontal loading factors on the elevator car, counterweights, and guide rails. Maximum deflections are also specified. Section 3041(d) requires emergency signal devices for earthquake emergency operations. Principally, the provisions stated in Section 3041(d) provide for elevator safety rather than elevator operability. CAC 24-7 only allows elevators, which are a vital circulation link in multistoried essential facilities, to operate from atop the car after the earthquake switches have been triggered. Once the individual shaft has been inspected by building personnel, the car can be returned to normal service. This is, however, a time consuming proposition. The trained personnel designated to inspect the vertical circulation systems may be better assigned to other more pressing tasks immediately after an earthquake. Would it not make more sense to design and qualify the elevator system to remain fully operational during and after an earthquake rather than relying on visual inspections after?

The elevator safety regulations are in direct response to the failures that occurred during the 1971 San Fernando earthquake. Fortunately, no personnel injuries were reported as a result of elevator failures. Had the earthquake occurred at a later time of the day, injuries could have been expected.

Elevators are a vital circulation link to many multistoried facilities, and hospitals are uppermost on this list. After the 1971 San Fernando earthquake, the Elevator Section of the State of California Department of Industrial Relations, Division of the Los Angeles City Department of Building and Safety requested reports on damage from the major elevator manufacturers. Ayres and Sun (1973) published the results of this compilation in "Nonstructural Damage," which appears in Volume 1, Part B of the N.O.A.A. publication *San Fernando, California, Earthquake of February 9, 1971.* An abridged version of Ayres and Sun's Table 1 is presented here as Table A1.14. Examination of this table shows that there were 674 counterweights displaced from their guide rails. Of these, 109 struck the elevator cars. Had the earthquake occurred during normal working hours, the chances are high that one of the collisions would have resulted in injured passengers.

Elevator machinery such as generators, motors, and electrical control panels also encountered major problems in the 1971 earthquake. CAC 24-7 revisions now require that all this equipment be anchored. This will significantly reduce the volume of equipment damaged in future earthquakes. Assuring that the electrical control panels will remain anchored does not guarantee that they will be able to perform satisfactorily after the earthquake. Accidental tripping of relays and false signaling could lead to control panel malfunctions, possibly putting the elevator out of commission. For critical elevators, this possibility cannot be tolerated.

TABLE A1.14.   Elevator Damage Summary (From State of California, Department of Industrial Relations, Division of Industrial Safety) as reproduced by J. Marx Ayres and Tseng-Yao Sun, *San Fernando, California Earthquake of February 9, 1971,* Vol. I, Part B, "Nonstructural Damage", L. M. Murphy, ed., N.O.A.A., Washington, D.C., 1973.

Item	Damage Description	Numbers
Buffers	Stinger damaged, guides damaged	2
Cables, compensating	Out of grooves or damaged	100
Cables, governor	Hung up, cut	20
Cables, hoisting	Damaged, jumped deflect sheave (sets)	7
Cables, traveling	Hung up or broken	7

TABLE A1.14. (Continued)

Item	Damage Description	Numbers
Cars	Out of guides, out of alignment	18
Cars	Steady rest loose	102
Controllers	Moved or damaged	5
Counterweights	Out of guide rails	674
Counterweights	Out of guide rails, damaged cars	109
Electrical power problems	Various causes	4
Electrical	Conduit loose	5
Generators, moved	Some damage	174
Generators	Burned out	5
Guide rails, car	Limits not operative, doors would not work	7
Guide rails, counterweights	Out of alignment, bent or broken	49
Guide rails, counterweights	Brackets broken or damaged	174
Hoistway doors	Off tracks, gibs out of sill, dragging out of alignment	22
Hoistway doors	Glass damaged	2
Hoistway walls	Bowed or hitting car	2
Hoistway walls	Severe cracks, loose plaster, and holes	50+
Hydraulic	Leaks in casing, plunger rubbing, out of alignment	8
Interlocks and car gate contact	Loose, broken	19
Leveling units	Damaged	12
Machine room	Floor plates broken, buckled	1
Miscellaneous	Annunciators off, brake rods out of core, light fixtures, broken selector tapes, fuse blowing, door operator	83
Motor, hoisting	Burned out, out of alignment, slipped rings	13
Pits flooded	Broken pipes, and so on	7
Pump units	Moved, reset leaks	1
Roller guides, counterweights	Broken or loose	286
Safeties set		22
Selector	Turned over	1
Sheaves, drive	Broken or cracked	3
Sheaves, counterweights	Broke loose, weight in pit moved	2
Shoes, guide	Broken	9
Victaulic fitting	Replaced	1

CAC 24-7 is the most comprehensive section of the *California Adminis-trative Code* reviewed for this book. Its only failure is that it does not provide an adequate avenue to reasonably assure that critical elevators will be able to operate after a significant earthquake. In fact, all the automatic switches and visual inspection requirements will most likely result in fewer operating elevators for essential facilities. Prior seismic qualification of elevator systems can alleviate the problem, thereby making more elevators available when they are most needed.

EARTHQUAKE RESISTANT DESIGN REQUIREMENTS FOR VA HOSPITAL FACILITIES, HANDBOOK H-08-8, VETERANS ADMINISTRATION (VA H-08-8)

The requirements imposed by VA H-08-8 are based on the Uniform Building Code (prior to UBC 1976). Table 2 of VA H-08-8 ties the force factor (C_p) for

TABLE A1.15.   Force Factor C_p for Parts or Portions of Buildings or Other Structures (Including Nonstructural Elements)

Parts or portions of buildings; tanks, towers, and equipment attached and ground supported; miscellaneous	Direction of force	Value of C_p A_{max} $\geq 0.15g$	$< 0.15g$
All walls, partitions and masonry, or concrete fences	Normal to flat surface	0.30	0.20
Cantilever parapet and other cantilever walls, except retaining walls	Normal to flat surface	1.00	1.00
Connections for prefabricated structural elements other than walls, with force applied at center of gravity of assembly	Any horizontal direction	0.50	0.30
Exterior and interior ornamentations and appendages	Any horizontal direction	1.00	1.00
Connections for exterior panels	Any horizontal direction	2.00	2.00
Ceilings, including integral lighting	Any horizontal direction	0.30	0.10
Hanging light fixtures	Any horizontal direction	1.00	0.50

Abridged from VA H-08-8, Table 2.

parts of buildings to the expected site acceleration value (A_{max}). Equation 3.9.1 in VA H-08-8 is shown below.

$$F_p = C_p W_p$$

VA H-08-8 does not modify C_p with importance factors, seismic zone factors, and so on as does UBC 1979. As is mentioned earlier, the site acceleration is accounted for by selecting a C_p value for those sites where A_{max} is expected to be less than $0.15g$ and another C_p value if A_{max} is greater than or equal to $0.15g$.

Pertinent portions of VA H-08-8, Table 2 have been reproduced here as Table A1.15. A comparison of the corresponding tables in UBC 1979 shows that VA H-08-8 is more explicit than UBC 1979 in that specific types of equipment are named. For A_{max} greater than or equal to $0.15g$, VA H-08-8 is decidedly more strict. The associated increase in C_p ranges from approximately 1.5 to 2 times.

Both VA H-08-8 and UBC 1979 recognize the importance of increasing C_p for flexible equipment mounted on the upper stories when the structure and equipment natural periods are coincident or nearly so. VA H-08-8 is similar to the *Uniform Building Code,* but it is slightly more definitive.

POST-EARTHQUAKE EMERGENCY UTILITY SERVICES AND ACCESS FACILITIES, VETERANS ADMINISTRATION CONSTRUCTION STANDARD CD-54 (VA CD-54)

VA CD-54 covers provisions required for electric power service, air conditioning systems, water service, sewer service, gas service, steam service, oxygen service, and site access facilities for Veterans Administration Hospitals. This regulation is limited to sites that are expected to have estimated damage of Modified Mercalli Intensity VIII or greater. The sites are expected to be capable of providing emergency services for a period of 4 days without any outside assistance.

EARTHQUAKE-RESISTIVE DESIGN OF NONSTRUCTURAL ELEMENTS OF BUILDINGS, VETERANS ADMINISTRATION CONSTRUCTION STANDARD CD-55 (VA CD-55)

The purpose of VA CD-55 is to establish the Veterans Administration's policy with regard to the design of the components of hospitals to resist damage during an earthquake. This includes architectural elements, electrical elements, and mechanical elements. VA CD-55 refers to VA H-08-8 for specific requirements and detailed instructions on equipment.

CITY OF LOS ANGELES: *LOS ANGELES MUNICIPAL CODE*

The *Los Angeles Municipal Code* is based on the *Uniform Building Code*.

CITY OF LOS ANGELES, *RECOMMENDED STANDARDS FOR SUSPENDED CEILING ASSEMBLIES*

This is a construction standard adopted by the Los Angeles City Department of Building and Safety for the installation of suspended ceilings. The following details are discussed in this standard.

- Vertical and horizontal static loads.
- Vertical support.
- Lateral support.
- Perimeter attachment.
- Lighting fixture and air diffuser support.

This construction standard requires the main ceiling runners to be able to withstand minimum compressive and tensile loads, specifies runner connection loads, and so forth. Manufacturers must submit test and/or calculation data to substantiate these requirements. Compression posts are not required by this standard.

CITY OF LOS ANGELES DEPARTMENT OF WATER AND POWER SPECIFICATION 9618, PART F, DIVISION F1, SECTION F1A, ARTICLE 27, ISSUED IN CONNECTION WITH ENERGY CONTROL PROJECT *COURTESY OF LADWP*

Excerpt

27. Earthquake Design: Earthquake design shall be as follows.
   a. **General:**   Where required in the detailed specifications, the equipment and its anchorage furnished and installed by the Contractor shall comply with the earthquake design requirements set forth hereinafter in this Article.

Design and analysis calculations or test results, including drawings, charts, and records, which show compliance with the seismic requirements of these specifications shall be signed by a California licensed Civil or Structural Engineer and shall be submitted for approval in accordance with Article 9 of this Section.

In order not to delay the work and to give the Engineer sufficient time to approve the earthquake design requirements, the Contractor shall correlate the data submittal time with the construction or installation starting date for the class of work as shown on the Progress Schedule called for in Subarticle 1d of Division E2.

b. Static Analysis: The equipment and its major components shall withstand the stresses caused by forces applied at the center of gravity of the equipment due to accelerations of $0.3g$ (g is the acceleration due to gravity) in any horizontal direction and $0.25g$ in either vertical direction, unless other values are specified. Horizontal and vertical inertial forces resulting from the base motion shall be considered to act simultaneously. These forces shall be applied to the operating, testing, or flooded weights whichever is greatest of the unit or part of the unit to be anchored or otherwise restrained.

Contractor's drawings referred to in Subarticle 27a of this Section shall include the following information.

(1) Installed weights and center of gravity of the equipment.

(2) Maximum base acceleration for which the equipment is designed.

c. Dynamic Analysis: The equipment furnished hereunder shall withstand the stresses caused by a vibratory ground motion with maximum accelerations of $0.5g$ (g is the acceleration due to gravity) in any horizontal direction. The vertical acceleration resulting from the ground motion shall be considered to be 80 percent of the horizontal acceleration and acting simultaneously with the horizontal acceleration in a direction which produces the most severe equipment stresses. The equipment shall continue to perform its intended function during and after such seismic stresses. The natural frequencies and damping of the equipment, mounted in the service configuration, with all accessories installed, shall be determined and the dynamic response shall be considered in its design.

The resulting internal stresses in a steel or aluminum component shall not exceed that component's yield strength.

Steel support frames shall be designed and fabricated in accordance with the AISC Specification for the Design, Fabrication, and Erection of Structural Steel for Building, as last revised.

Aluminum support frames shall be designed and fabricated in accordance with Chapter 28 of the Uniform Building Code.

The resulting internal stresses in a ceramic component shall not exceed 50 percent of that component's ultimate mechanical strength.

Seismic-withstand capability shall be demonstrated either by test or analysis in accordance with one of the following methods or a combination of the methods described hereinafter. Any damping values used in the analytical methods greater than 2 percent of the critical damping shall be supported by test records as proof of the damping value.

(1) Method A—Static Analysis of Rigid Equipment: Method A may be used only where the natural frequencies of the equipment exceed 30 hertz. The equipment shall be designed to withstand inertial stresses resulting from an acceleration of the equipment base equal to the maximum ground motion accelerations.

(2) Method B—Modal Analysis of Flexible Equipment: Using Method B, the equipment shall be modeled as a series of discreet mass points connected by mass-free members. Sufficient mass points shall be used to insure an adequate representation. The model shall represent the equipment as it will

be mounted in service. The resulting system shall be dynamically analyzed using "modal spectrum analysis," as described in the book, "Introduction to Structural Dynamics," by J. M. Biggs, published by McGraw-Hill in 1964. The maximum modal response shall be determined using the response spectra shown on the Seismic Design Response Spectra* on Page F1A-18. The total response shall be determined by combining the modal response by the square root of the sum of the squares technique.

(3) Method C—Time-History Analysis of Flexible Equipment: Using Method C, equipment shall be modeled as a single or multiple degree-of-freedom system and dynamically analyzed by the Time-History Method. For the Time-History Method, the ground motion acceleration time histories will be furnished on request.

(4) Method D—Test: Tests shall be performed by subjecting the equipment furnished under these specifications to vibratory motion which conservatively simulates that to be seen at the equipment mounting. The equipment shall be mounted during the test in a manner that simulates the intended service mounting. Tests shall be conducted in two phases as follows:

(a) Phase I—Resonant Frequency Search: A low-amplitude frequency search shall be conducted at a rate not greater than two octaves per minute in the range from 0.1 to 33 hertz to determine the regions of the resonance in the three axes. The test equipment base acceleration input during the sweep shall be in the range from $0.1g$ to $0.2g$.

(b) Phase II—Full Scale Test: Tests shall be conducted at the resonant frequencies determined in Phase I with amplitudes of base acceleration equal to $0.5g$ in the horizontal directions, coincident with the equipment axes, and 80 percent of the horizontal acceleration in the vertical direction. The tests may be in one axis at a time if it can be shown that no significant coupling exists in the equipment between the horizontal and vertical axes to give additive responses in the unexcited axis or if an adjusted input is applied which will account for the additive responses caused by the coupling. Either of the two following methods of application of the base acceleration may be used.

(aa) Continuous Test: A continuous sinusoidal motion shall be applied for a duration of 15 cycles of each resonant frequency.

(bb) Sine Beat Test: A sinusoidal beat motion consisting of a sinusoid of the equipment resonant frequency modulated by a lower frequency sinusoid which provides at least 10 cycles of resonant frequency per beat. There shall be a minimum of 5 such bursts of resonant frequency within a period ranging from 60 seconds to 150 seconds and the pause between bursts shall be long enough so that there will be no significant superposition of motion. This testing method shall be as described in the publication "IEEE Guide for Seismic Qualification of Class 1 Electric Equipment for Nuclear Power Generating Stations," IEEE Standard 344, as last revised.

(5) Other Methods: Any other method meeting the requirements of "IEEE Guide for Seismic Qualification of Class 1 Electric Equipment

---

*This response spectra has been omitted from this appendix.

for Nuclear Power Generating Stations," IEEE Standard 344, as last revised, will satisfy the requirements of these specifications.

Contractor's drawings referred to in Subarticle 27a of this Section shall include the following information:

(1) Natural frequencies and damping.

(2) Maximum acceleration due to ground motion for which the equipment is designed.

(3) Center of gravity and installed weights of the equipment.

(4) All dimensions and descriptions, such as cross sections and other mechanical properties, necessary to perform stress analysis of the equipment and supporting structure.

(5) Model of equipment used.

Anchorage of equipment shall comply with Subarticle 27d of this Section.

   d. **Anchorage of Equipment:** Equipment shall be anchored to structural load carrying building components. Anchoring devices and bolts shall be designed to resist earthquake motion accelerations of 0.5g horizontal and 0.4g vertical (g is the acceleration due to gravity) acting simultaneously. Stresses induced by this motion shall be working level stresses with no allowable increase for seismic, determined by using static analysis.

Contractor's drawings referred to in Subarticle 27a of this Section shall show anchorage details including anchor bolt patterns and bolt sizes.

Discussion

Since the devastating results of the 1971 San Fernando earthquake, the Los Angeles Department of Water and Power (LADWP) has been pioneering in the field of earthquake qualification for its facilities. This is true for both their newly constructed and existing (retrofitting or backfitting) facilities. Their earthquake design specification for the Energy Control Center project is just one example of their dedication to improving the Department of Water and Power's margin of safety in a major earthquake. The LADWP facilities fit into the category of what has been termed "life line earthquake engineering." These facilities are all roughly synonymous with "essential facilities" as the term is used in this text.

This LADWP specification is short and yet concise. Article 27, Paragraph C, under dynamic analysis states that

   . . . the equipment shall continue to perform its intended function during and after such seismic stresses. . . .

All mathematical analyses are limited in demonstrating operability after an earthquake. They cannot prove operability wherever equipment is likely to suffer a functional type of failure. If equipment only needs to remain anchored to remain functional, then analysis can demonstrate the operability requirement of Paragraph C. This can only be the case where subcomponents of the individual piece of equipment are also not likely to fail as a result of dynamic motions.

LADWP's use of the static analysis approach for rigid equipment (where the natural frequency exceeds 30 hertz) is generally acceptable. It does not, however, specifically mention internal subcomponents that may have lower natural frequencies and be more susceptible to dynamic motions. The modal analysis suggested in method B is the most widely used and accepted method of dynamic analysis. The square root-sum-of-the-squares is a method that prevents overly conservative design. Care must be exercised in selecting a response spectrum applicable for the individual equipment location. Method C, the time-history analysis, is less often used. Although it leads to more precise solutions to the problem, it is more complex than the other methods and generally more costly.

LADWP's seismic specification provides for seismic testing. It does not, however, detail when and for which equipment a test should be employed. The seismic test is generally used in cases where it is necessary to prove operability, in situations requiring proof and fragility tests (generic applications), or in cases where equipment is too complex or modeling assumptions too gross to analyze adequately. The continuous sine and the sine beat tests cannot adequately envelop the required response spectrum (RRS) nor can it generally prove equipment operability. By deferring to IEEE 344, LADWP has, however, given independent testing laboratories a vehicle by which they can adequately test a piece of equipment. The laboratory can select (with owner approvals) which waveform and type of test is best suited for each application. This is a decision that the laboratory is extensively qualified to make.

In general, this Los Angeles Department of Water and Power specification is quite useful and is a large step forward for facilities and their equipment qualification programs for the earthquake environment.

FOREIGN SEISMIC CODES

BULGARIA: *BULGARIAN CODE FOR BUILDING IN EARTHQUAKE REGIONS, 1964*

The Bulgarian code uses a static coefficient method similar to that of the *Uniform Building Code*. The code requires electric conduits to be capable of elongating as a result of earthquake motions.

CANADA: *NATIONAL BUILDING CODE OF CANADA 1975*

The Canadian code is based on parameters similar to those on which *Uniform Building Code* was based prior to 1976.

EL SALVADOR: *REGULATIONS FOR SEISMIC DESIGN REPUBLIC OF EL SALVADOR, C.A. 1966*

The El Salvador code simply requires that equipment be fixed to the structural resisting elements of the building and be installed so as to accommodate the building's displacement.

FRANCE: *BUILDING REGULATIONS FOR SEISMIC AREAS 1967*

The French code discussion limits itself to nonbearing walls, partitions, and small freestanding elements. These items are qualified by a static coefficient calculation using a modified acceleration value that is dependent on the building importance and the seismic zone for which the building is to be built.

ISRAEL: *PROPOSED ISRAEL STANDARD LOADS IN BUILDINGS: EARTHQUAKES*

The Israeli code relies on the static coefficient method for seismic qualification of equipment.

NEW ZEALAND: *NEW ZEALAND STANDARD, NZS 4203: 1976*

The New Zealand code is based on the static coefficient method of design for nonstructural elements of buildings. It is, however, the most complete foreign code reviewed with respect to equipment.
    Items such as the following are covered:

- Containers and contents.
- Furnaces, boilers, transformers, switchgear, shelving.
- Horizontal cantilevers.
- Lift machinery and standby equipment (emergency power supplies).
- Ornamentations.
- Suspended ceilings.
- Tanks, towers and stacks.
- Vertical cantilevers.

The New Zealand 1976 code assigns higher seismic design coefficients (C_p) to the individual items than does the similar UBC 1976. The C_p values for the New Zealand code are also much more explicit for the various possible cases than the UBC 1976.

PERU: *PERUVIAN STANDARDS FOR ANTISEISMIC DESIGN, 1968*

The Peruvian code is a simple static coefficient method of seismic analysis for equipment and requires only anchorage.

RUMANIA: *EARTHQUAKE REGULATIONS RUMANIAN PEOPLE'S REPUBLIC*

The Rumanian code is a simple static coefficient method of seismic analysis for equipment and requires only anchorage.

VENEZUELA: *PROVISIONAL STANDARD FOR EARTHQUAKE-RESISTANT STRUCTURE 1967*

The Venezuelan code is a simple static coefficient method of seismic analysis for equipment and requires only anchorage.

CONCLUDING REMARKS ON CODES AND STANDARDS

For centuries, buildings have been designed for the static environment. Any loads imposed on them were assumed to occur over a long period of time and the resultant reactions were easily calculated; but then came the earthquake consideration. Building loads are often applied over a very short time span, are repetitive, have more than one direction of occurrence, and have resulted in reactions that are slightly more difficult to predict.

It has taken two major earthquakes (Alaska 1964 and San Fernando 1971) affecting urban areas within less than a decade to alert building code authors to the importance of essential facilities such as hospitals and communication centers and the necessity of their operability immediately after an earthquake. The mere mention of operability in the building code, however, does not guarantee essential facility operation. Building codes and design specifications that spell out mechanisms for adequate seismic qualification are urgently needed. Currently, facilities are designed and built so that the building structure itself is quite competent. The building equipment, however, is commonly neglected. The scenario that we can expect from this condition after future destructive earthquakes is relatively undamaged buildings, with much of the equipment necessary for the building to function piled on the floor in an unusable mess.

The *Uniform Building Code* recognizes essential facilities and some types of building equipment (base ancorage required). The seismic qualification approach taken by the *Uniform Building Code* is, however, not stringent enough to suggest with any degree of confidence facility operation after a significant earthquake. For example, most facilities will experience a large amount of downtime as a result of overturned equipment (both stationary and portable), items spilled from shelves, and disarrayed office furniture.

Although not yet adopted, the most comprehensive general use seismic code to date is the Applied Technology Council *Tentative Provisions for the Development of Seismic Regulations for Buildings*. This code contains provisions for quite an array of architectural, mechanical, and electrical components. It considers more than simple base anchorage (dynamic characteristics, mounting systems, component geometry, etc.) and provides for qualification procedures previously not considered in seismic codes. These comprehensive qualification procedures include seismic shake table testing and dynamic analysis. This alone is a significant step in increasing the margin of safety for essential facilities.

The *State of California, California Administrative Code,* "Elevator Safety Regulations" were rewritten primarily as a result of the 1971 San Fernando earthquake. The effect of this rewrite, however, has been to guarantee that elevators will be placed out of commission immediately after a significant earthquake. The technology exists to design elevators to remain operational unless a collision is imminent. Operational elevators will obviously better serve essential facilities than those that are shut down.

Items that have not been addressed by any of the aforementioned codes include proof of critical equipment operability, qualification of owner supplied equipment, consideration of counter top and shelved items, as well as general office furniture. The codes also do not consider the possibility of equipment qualification by prior qualification (similarity to equipment previously installed) and designer judgment. Simple, imaginative architectural detailing can solve many seismic requirements.

Most importantly, however, with the exception of the California elevator code, no consideration is given to existing essential facilities. Most of the facilities currently in use were built prior to the 1976 edition of the *Uniform Building Code*. This simply means little or no qualification consideration has been given to equipment. Interested readers need only visit a handful of "older" essential facilities to realize the importance of providing for backfitting programs by any future code.

Appendix 2. Equipment Manufacturers

- Table A2.1 Access Floor Systems
- Table A2.2 Air Handling Systems
- Table A2.3 Elevator Systems
- Table A2.4 Emergency Power Supply Systems
- Table A2.5 Fire Protection Systems
- Table A2.6 Kitchen Systems
- Table A2.7 Lighting Systems
- Table A2.8 Medical Systems
- Table A2.9 Suspended Ceiling Systems
- Table A2.10 Water Systems
- Table A2.11 Miscellaneous Equipment

This appendix has been included to assist the design team in selecting equipment for the various systems. Manufacturers have been noted by system and equipment type in Tables A2.1 through A2.11. Because of the ever evolving equipment designs, no attempt has been made to judge the various manufacturers or their products with respect to earthquake survivability. For some of the equipment considerable effort has been put into earthquake qualification programs. For others little or none has been made. Some equipment items, however, do have effective inherent resistance to earthquake induced failure from other design features. This often results in equipment that survives an earthquake even though no specific attention has been paid to it in the design, installation, or manufacturing process.

The design team must review each piece of equipment for each application and weigh the individual merits. As time progresses, and as design teams become more aware of the qualification processes with each successive project, the equipment selection task will become more straightforward. If adequate notes are kept, the design team will be able to apply the "qualified by prior experience" approach. This will result in an obvious savings to the facility owners both in time expended and dollars.

TABLE A2.1. Access Floor Systems

Equipment	Manufacturer	Remarks
Access Floor Inter-loc™	Liskey Architectural Mfg., Inc. P. O. Box 8748 Baltimore, MD 21240 (301) 796-3300	Interlocking floor panels
Access floors	Floating Floors, Inc. 6955 Wales Road Toledo, OH 43619 (419) 666-8750	Grid and gridless floors
Access floors	Donn Corporation 1000 Crocker Road Westlake, OH 44145 (216) 871-1000	
Access floors	Mult-a-frame Corp. Mult-a-floor Division 366 South Boulevard East Pontiac, MI 48053 (313) 338-9271	
Access floors	Tate Architectural Products, Inc. Montevideo Road Jessup, MD 20794 (301) 799-4200	
Access floors	Westinghouse Access Flooring Architectural Systems Division 4300 36th Street, S.E. Grand Rapids, MI 49508 (616) 949-1050	
Seismic floors	Ohbayashi-Gumi, Ltd. Tokyo, Japan	Manufactured for the seismic environment

TABLE A2.2. Air Handling Systems

Equipment	Manufacturer
Air conditioners, heat pumps, heating/ cooling units, condensing units, air handlers, oil fired furnaces, gas fired furnaces, and unit heaters	BDP Company Day & Night Payne Bryant Phoenix, AZ (602) 269-2351
Air conditioning, heating, chillers, condensors, compressors, coolers, fans, burners, and pumps	Dunham-Bush, Inc. 175 South Street West Hartford, CT 06110

412

TABLE A2.2. (Continued)

Equipment	Manufacturer
Air conditioning	Temp Master Corp. 1222 Ozark Street North Kansas City, MO 64116 (816) 421-0723
Air conditioning equipment, condensing units, air handlers, heat pumps, and warm air furnaces	General Electric Central Air Conditioning 4421 Bishop Lane Louisville, KY 40218 (502) 452-3165
Air curtains	Dynaforce Corp. 195 Sweet Hollow Road Old Bethpage, NY 11804 (516) 420-8787
Air handling, air curtains, and refrigeration	The King Company 1037 21st Avenue Owatonna, MN 55060 (507) 451-3770
Air handling, makeup units, fans, pumps, and heat recovery	Buffalo Forge Company 490 Broadway Buffalo, NY 14204
Air handling	Webco Engineering Co. 2325 E. Trafficway Springfield, MO 65802 (417) 866-7231
Air handling, air conditioning, and heating equipment	Table Air Conditioning 3600 Pammel Creek Road La Crosse, WI 54601
Chillers, compressors, roof top units, heat pumps, air handlers, condensers, fans, heaters, and computer room air conditioners	Trane Air Conditioning Commercial Air Conditioning Division La Crosse, WI 54601
Duct heaters, and hot water boilers	York Heating and Air Conditioning 425 Hanley Industrial Ct. St. Louis, MO 63144 (314) 644-4300
Fans	Twin City Fan and Blower Company 550 Kasota Ave. S.E. Minneapolis, MN 55414 (612) 331-4104
Fans	Acme Engineering and Manufacturing Corp. 1820 N. York Muskogee, OK 74401 (918) 682-7791
Flexible air ducts	The Wiremold Co. 4792 Gregg Road Pico Rivera, CA 90660 (213) 669-0558

TABLE A2.2. (Continued)

Equipment	Manufacturer
Flexible lines—air conditioning	Aeroquip Industrial Division Jackson, MI 49203
Gas heaters—industrial	Detroit Radiant Products 1297 Terminal Avenue Detroit, MI 48214 (313) 823-1074
Heat pumps, heating units, cooling units, air conditioning, air cleaners, humidifiers, condensing units, chillers, air handlers, fans, air distribution boxes, and air terminals	Carrier Air Conditioning Corp. Carrier Parkway Syracuse, NY 13221 (315) 432-7492 (315) 424-4994
Heating, ventilating, air conditioning, and makeup air	E. K. Campbell 1809 Manchester Trafficway Kansas City, MO 64126 (816) 483-3300
Heating, ventilating, and air conditioning equipment	Lennox Industries 200 South 12th Ave. Marshalltown, IA 50158 (515) 754-4011
Solar equipment, air conditioning, ventilators, fans, and packed equipment	American Air Filter 215 Central Avenue Louisville, KY 40277 (502) 451-2183
Vibration isolation	(See Emergency Power Supply Systems, Table A2.4)

TABLE A2.3. Elevator Systems

Equipment	Manufacturer
All elevator equipment	Otis Elevator Company One Farm Springs Farmington, CT 06032
All elevator equipment	Dovor Corporation Elevator Division P. O. Box 2177 Memphis, TN 38101
All elevator equipment	Montgomery Elevator Co. Moline, IL 61265 (309) 764-6771

TABLE A2.3. (Continued)

Equipment	Manufacturer
All elevator equipment	Westinghouse Elevator Co. 21 Bleeker Street Millburn, NJ 07041
Hydraulic elevators	Esco Elevators, Inc. P. O. Box 445 Fort Worth, TX 76101 (817) 332-7655

TABLE A2.4. Emergency Power Supply Systems

Equipment	Manufacturer	Remarks
Emergency power supplies	Caterpillar P. O. Box 351 Riverside, CA 92502 (714) 686-4560	
Emergency power supplies	Equipment Service Company (ESCO) P.O. Box 1307 Long Beach, CA 90801 (213) 426-0311	
Emergency power supplies	Solar Turbines International 2200 Pacific Highway San Diego, CA 29138 (714) 238-5500	Gas turbine plus all periferals
Vibration isolators and snubbers	California Dynamics Corp. 5572 Alhambra Ave. Los Angeles, CA 90032 (213) 223-7882	
Vibration isolators— lockouts	Consolidated Kinetics Corp. 249 Fornof Columbus, OH 43207 (614) 443-7621	
Vibration isolators and snubbers	Mason Industries, Inc. 3335 E. Pico Blvd. Los Angeles, CA 90023 (213) 263-9557	

TABLE A2.5.   Fire Protection Systems

Equipment	Manufacturer	Remarks
Extinguisher bottles, extinguisher brackets, extinguisher cabinets, hose cabinets, hose racks, and safety cabinets	J. L. Industries 4450 W. 78th Street Bloomington, MN 55435 (612) 835-6850	Extinguishers, cabinets, hoses, and so on
Extinguisher bottles, extinguisher brackets, extinguisher cabinets, hose cabinets, hose racks, and safety cabinets	Larsen's Manufacturing Co. 7421 Commerce Lane, N.E. Minneapolis, MN 55432 (612) 571-1181	Extinguishers, cabinets, hoses, and so on
Halon fire units and extinguisher bottles	Kidde-Sentinel 675 Main Street Belleville, NJ 07109 (201) 759-5000	
Sprinklers, wet chemical extinguishers, and halon fire units	Automatic Sprinkler Corp. of America P.O. Box 180-S Cleveland, OH 44147 (216) 526-9900	
Sprinklers, alarms, detectors, and communications	Grinnell Fire Protection Systems Company, Inc. 10 Dorrance Street Providence, RI 02903	

TABLE A2.6.   Kitchen Systems

Equipment	Manufacturer
Back bar equipment, food and dish carts, fryers, ranges and ovens, broilers, griddles, steam cookers, kettles, beverage dispensers, refrigeration, dishwashers, serving counters, sinks, ventilators	Alco Food Service Equipment Co. 8181 N.W. 36th Street Miami, FL 33166 (305) 592-3660
Beverage equipment, refrigerators, and freezers	Nor-Lake Inc. Second & Elm Hudson, WI 54016 (715) 386-2323
Cafeteria counters, food warmers, tables, portable servers, chef's units	Duke Manufacturing Co. P. O. Box 5426 St. Louis, MO 63147 (314) 231-1130

TABLE A2.6. (Continued)

Equipment	Manufacturer
Exhaust hoods	Gaylord Industries, Inc. P. O. Box 558 Wilsonville, OR 97070 (503) 682-3801
Food scales, freezers refrigerators choppers cutters saws mixers peelers slicers disposers dishwashers	Hobart Troy, OH 45374
Ice machines	Manitowoc Equipment Works 2110 South 26th Street Manitowac, WI 54220 (414) 682-0161
Ice machines	Scotsman Queen Products Division King-Seeley Thermos Co. 505 Front Street Albert Lea, MN 56007
Ice machines	Mile High Equipment Co. Ice-o-matic 11100 East 45th Avenue Denver, CO 80239 (303) 371-3737
Icc machines	Uniflow Manufacturing Co. Kold-Draft Division East Lake Road Erie, PA 16512 (814) 453-6761
Ice makers	Frigidaire 300 Taylor Street Dayton, OH 45442 (513) 445-9025
Metal sinks	Metal Masters Food Service Equipment Co. 655 Glenwood Avenue West Smyrna, DE 19977 (302) 653-5871
Pizza ovens, broilers, bake and roast ovens, slicers, deep fryers	Connolloy Roll-a-grill 12 First Street Pelham, NY 10803 (914) 738-4333
Refrigerators	Balley Case and Cooler, Inc. Balley, PA 19503
Wheeled food cabinets, serving counters, dish dispensers, automatic fryers, steamers, roasting ovens	Cres-cor 12711 Taft Avenue Cleveland, OH 44108

TABLE A2.7. Lighting Systems

Equipment	Manufacturer
Emergency lights	Carpenter Lighting Systems 706 Forrest Street Charlottesville, VA 22902 (804) 977-8050

TABLE A2.8. Medical Systems

Equipment	Manufacturer	Remarks
Autopsy tables, dissecting tables, mortuary casework, refrigerators, racks, and cadaver lifts	Lipshaw Manufacturing 7446 Central Avenue Detroit, MI 48210	
Casework	St. Charles Manufacturing Co. St. Charles, IL 60174 (312) 584-3800	
Fume hoods, casework, counters, and cabinets	Lab Fabricators Company 1800 East 47th Street Cleveland, OH 44103 (216) 431-5444	
Gas sterilizers, aeration cabinets, drying cabinets, and stacking racks	Steri-vac Infection Control Sterilizers Medical Products Division/3M 3M Center St. Paul, MI 55101	
Hospital refrigerators, bloodbanks (wall-mounted, floor-models, wheeled models), Blood plasma freezers, pharmaceutical refrigerators, morgue refrigerators, dissecting sinks, mobile tables, and autopsy tables	The Jewett Refrigerator Company, Inc. 2 Letchworth Street Buffalo, NY 14213 (716) 881-0030	Bloodbanks, refrigerators, and so on
Laboratory casework, laboratory shelving, and fume hoods	Duralab Equipment Corp. 107-23 Farragut Road Brooklyn, NY 11236 (212) 649-9600	
Laboratory casework, and counter tops	Johns-Manville Ken-Caryl Ranch Denver, CO 80217 (303) 979-1000	

Equipment	Manufacturer	Remarks
Laboratory furniture, casework, and fume hoods	Fisher Scientific Co. Contempra Furniture Division 1410 Wayne Avenue Indiana, PA 15701 (800) 245-6897	
Medical gas equipment, compressors, alarms, pumps, bulk O_2, gas bottles, and service units	Chemetron Corporation Medical Products Division 1801 Lilly Avenue St. Louis, MO 63110 (314) 771-2400	
Medical storage modules, sinks, cabinets, shelves, carts, and medication centers	Macbick Murray Hill, NJ 07974	
Medical wall panels ICU CCU Surgery X-Ray Hemodialysis	Modular Services Company 110 Northeast 38th Terrace Oklahoma City, OK 73105 (405) 521-9923	
Medical walls and monitor mounts	Inter Royal Corp. 1 Park Avenue New York, NY 10016	
Mobile food cabinets	Crescent Metal Products, Inc. 12711 Taft Cleveland, OH 44108 (216) 851-6800	
Modular hospital walls	Square "D" Company 3300 Medalist Drive Oshkosh, WI 54901	Oxygen, electricity, monitors, and so on are provided
Nurses stations, built-in casework	SCI P. O. Box 11017 Chicago, IL 60611 (312) 337-3414	
Patient monitoring equipment	Mennen-Greatbatch Electronics, Inc. 10123 Main Street Clarence, NY 14031 (716) 759-6921	
Patient tables, portable carts, sterilizers, surgical lights, washers, and clinical equipment	Sybron/Ritter P. O. Box 848 Rochester, NY 14603 (716) 436-6600	

TABLE A2.8. (Continued)

Equipment	Manufacturer	Remarks
Pharmacy casework	Grand Rapids Sectional Equipment Company P. O. Box 6306 Grand Rapids, MI 49506 (616) 243-4963	Wood casework
Portable hospital carts	Blickman Health Industries, Inc. 20-21 Wagaraw Road Fair Lawn, NJ 07410 (201) 423-3660	
Respiratory trees, wheeled	Devilbiss Company Medical Products Division 300 Phillips Avenue Toledo, OH 43692 (419) 470-2169	
Sterilizers, autoclaves, steam generators, patient tables, kitchen equipment, task lighting, casework, storage systems, and warming cabinets	American Sterilizer Co. 12990 Branford Street Unit 6B Arleta, CA 91331	

TABLE A2.9. Suspended Ceiling Systems

Equipment	Manufacturer	Remarks
Aluminum ceiling grids	Gordon 2000 Creswell Avenue S. Hreveport, LA 71104 (318) 424-0919	
Aluminum ceiling grids	Howmet Aluminum Corp. P. O. Box 40 Magnolia, AR 71753 (501) 234-4260	
Aluminum ceiling grids	National Rolling Mills Co. P. O. Box 622 Paoli, PA 19301 (215) 644-6700	
Aluminum ceiling grids	Roblin Building Products 1971 Abbott Road Buffalo, NY 14218 (716) 825-6650	

Equipment	Manufacturer	Remarks
Aluminum panel acoustical ceilings	Simplex Ceiling Corp. 663 Fifth Avenue New York, NY 10022 (212) 349-1890	Bolted, pinned, and snap fitted connections, compression posts (hangers)
Ceiling boards and panels	Owens-Corning Fiberglas Corp. Interiors Marketing Division Fiberglas Tower Toledo, OH 43659 (419) 248-8101	
Ceiling suspension systems	Haertel 11550 W. King Street Franklin Park, IL 60131 (312) 455-0727	
Decorative ceilings	Zero Dec Box H Newtown Square, PA 19073 (215) 356-2712	Fits aluminum ceiling grid
"Floating" suspended ceilings, ceiling framing	Alcan Building Products 280 North Park Avenue Warren, OH 44482	Suspended by wire hangers and seismic clips
Lay-in metal panels, acoustical metal ceilings	Steel Ceilings, Inc. 500 N. Third Street Coshocton, OH 43812 (614) 622-4655	Some use "pressure fitted" rod hangers, others wire supported
Lay-in panel ceilings, ceiling framing	Alpro Acoustics Division Structural Systems Corporation P. O. Box 50070 New Orleans, LA 70150 (504) 522-8656	Industrial and commercial applications
Lay-in panels, mirrored ceilings, air distribution tiles	United States Gypsum 101 South Walker Drive Chicago, IL 60606	
Lay-in panels	Chicago Metallic Corp. 5501 Downey Vernon, CA 90058 (213) 589-5771	
Lay-in panels and tile ceilings, ceiling framing	Gold Bond Building Products 2001 Rexford Road Charlotte, NC 28211	

TABLE A2.9. (Continued)

Equipment	Manufacturer	Remarks
Luminaire, lay-in panels, and tile ceilings; ceiling framing, suspended and integrated ceilings	Armstrong Architectural Ceiling Systems 9700 Flair Drive El Monte, CA 91731 (213) 585-5421	
Suspended and integrated ceilings	Celotex Corporation Western Acoustical Division P. O. Box 31178 Lincoln Heights Station Los Angeles, CA (213) 223-0911	
Suspended and integrated ceilings	Chicago Metallic Corp. 5959 E. Telegraph Road Los Angeles, CA 90040 (213) 726-6741	
Suspended and integrated ceilings	Conwed Corporation Ceiling Products Division 332 Minnesota Street St. Paul, MN 55101 (612) 221-1184	
Suspended and integrated ceilings	Ceiling Dynamics 1845 Belcroft Avenue S. El Monte, CA 91733 (213) 579-2652	Lay-in panels, lighting, and air distribution
T-bar ceilings	Amico P. O. Box 3928 Birmingham, AL 35208 (205) 787-2611	

TABLE A2.10. Water Systems

Equipment	Manufacturer
Boilers—flexible tube	Bryan Steam Corporation P. O. Box 27 Peru, IN 46970
Boilers—oil and gas fired	The Peerless Heater Co. Spring & Shaeffer Streets Boyertown, PA 19512
Boilers—oil and gas fired	H. B. Smith Co., Inc. 57 Main Street Westfield, MA 01085

TABLE A2.10. (Continued)

Equipment	Manufacturer
Centrifugal pumps, heat exchangers, refrigeration components, plumbing products, valves and fittings, boiler feed units, deaerators, condensate transfer units, and backflow preventers	Bell & Gossett 8200 N. Austin Avenue Morton Grove, IL 60053
Centrifugal pumps, heat exchangers, air separators, and pumps	Thrush Products, Inc. P. O. Box 228 Peru, IN 46970
Drinking fountains	EBCO Manufacturing Co. Oasis Drinking Water Fountains 265 North Hamilton Road Columbus, OH 43213 (614) 861-1350
Drinking fountains	Halsey Taylor P. O. Box 618 Bensenville, IL 60106 (312) 766-8000
Expansion joints (for piping)	Holz Rubber Company 1129 South Sacramento Street Lodi, CA 95240 (209) 368-7171
Flexible pump connectors, vibration absorbers (flexible hose)	Packless Industries P. O. Box 8799 Waco, TX 76710 (817) 776-7700
Humidifiers	Herrmidifier 1770 Hempstead Road Lancaster, PA 17604
Humidifiers	Walton One Carol Place Moonachi, NJ 07074 (201) 641-5700
Safety showers	Fisher Scientific Co. Contempra Furniture Division 1410 Wayne Avenue Indiana, PA 15701 (800) 245-6897
Water coolers	Elkay Manufacturing Co. 2700 S. 17th Avenue Broadview, IL 60153 (312) 681-1880
Water coolers	Halsey Taylor Route 75 Freeport, IL 61032 (815) 235-0066

TABLE A2.10. (Continued)

Equipment	Manufacturer
Water coolers	Oasis Water Coolers 265 N. Hamilton Road Columbus, OH 43213 (614) 861-1350
Water coolers	White-Westinghouse Commercial Products 246 E. 4th Street Mansfield, OH 44902

TABLE A2.11. Miscellaneous Equipment

Equipment	Manufacturer
Book shelving	Adjustable Steel Products Co. Division of Decision Systems Inc. 122 E. 42nd Street New York, NY 10017 (212) 986-9640
Book shelving	Art Metal USA, Inc. Aetnastak Division 300 Passaic Street Newark, NJ 07104 (201) 485-5310
Fixed shelving	E-Z Shelving, Inc. P. O. Box 5218 Kansas City, KS 66119 (913) 281-1112
High-rise shelving (for bulk storage)	Lyon Metal Products, Inc. 2933 Railroad Avenue Aurora, IL 60507
Library furniture, and book shelving	Library Bureau 801 Park Avenue Herkimer, NY 13350 (315) 866-1330
Merchandizing fixtures and storage shelves	Reeve Store Equipment 9131 E. Bermudez Street Pico Rivera, CA 90660 (213) 723-4791
Personnel lockers	Penco Products Inc. Oaks, PA 19456 (215) 666-0500

Equipment	Manufacturer
Personnel lockers	Republic Steel Industrial Products Division 1038 Belden Avenue, N.E. Canton, OH 44705 (216) 493-2795
Personnel lockers	De Brough Manufacturing Co. 9300 James Avenue South Minneapolis, MN 55431 (612) 884-5255
Personnel lockers	American Locker Security Systems, Inc. Box 489 Jamestown, NY 14701 (716) 487-0161
Personnel lockers	Interior Steel Equipment Co. 2352 East 69th Street Cleveland, OH 44104 (216) 881-0100
Personnel lockers—(mobile units)	List Industries, Inc. 21470 S. Main Street Matteson, IL 60443 (312) 481-7680
Personnel lockers	Lyon Metal Products, Inc. 18955 E. Railroad Street Industry, CA 91749
Pipe hangers and supports	Grinnel Company, Inc. 260 W. Exchange Street Providence, RI 02901
Shelving, track storage cabinets, casework, and carts	MEG 100 Bidwell Road South Windsor, CT 06074 (203) 289-8267
Shelving	HC Products Co. P. O. Box 68 Princeville, IL 61559 (309) 385-4334
Track storage cabinets, divider type storage, book shelving, and library furniture	Estey Corporation Drawer E Redbank, NJ 07701 (201) 542-5000
Track storage cabinets	Lundia, Myers Industries, Inc. 600 Capitol Way Jacksonville, IL 62650 (217) 243-8585

TABLE A2.11. (Continued)

Equipment	Manufacturer
Track storage cabinets, carriages for existing shelving, and book shelving	Spacesaver Corp. 1450 Janesville Avenue Ft. Atkinson, WI 53538 (414) 563-6362
Vending machines	Rowe International, Inc. 75 Troy Hills Road Whippany, NJ 07981 (201) 887-0400
Wardrobes, chests, builtin casework, and study carrels	SCI P. O. Box 11017 Chicago, IL 60611 (312) 337-3414

Appendix 3. *Guidelines for Seismic Restraints of Mechanical Systems* **and** *Guidelines for Seismic Restraints of Kitchen Equipment*

Chapter 4 contains diagrammatic design details for many types of equipment. This appendix has been included to complement Chapter 4 with guidelines that have been prepared by Hillman, Biddison & Loevenguth for the Sheet Metal Industry Fund of Los Angeles. They are titled *Guidelines for Seismic Restraints of Mechanical Systems* and *Guidelines for Seismic Restraints of Kitchen Equipment*. These guidelines are currently in the process of being republished in a new and expanded format by the Sheet Metal Industry Fund of Los Angeles. The reader will find it useful to obtain a copy of the new guidelines when they become available.

Reproduced with the permission of the Sheet Metal Industry Fund of Los Angeles.

GUIDELINES FOR SEISMIC RESTRAINTS OF MECHANICAL SYSTEMS (SHEETS 1 THRU 18)

PREPARED FOR:
SHEET METAL INDUSTRY FUND OF LOS ANGELES

PREPARED BY:
HILLMAN, BIDDISON & LOEVENGUTH STRUCTURAL ENGINEERS

JC Loevenguth S808

APPROVED BY:

OFFICE OF THE STATE ARCHITECT
STRUCTURAL SAFETY SECTION

APPROVED March 30 1976

PER *DK Jenkins* n.

PRINCIPAL STRUCTURAL ENGINEER

DATE APPROVED   MARCH 1976

REVISIONS AND/OR CHANGES TO ANY OF THE SHEETS 1 thru 18 WILL REQUIRE NEW APPROVAL BY O.S.A.

SMACNA

GUIDELINES FOR SEISMIC RESTRAINTS
OF
MECHANICAL SYSTEMS
© SMIF

SHEET METAL INDUSTRY FUND OF LOS ANGELES

401 SHATTO PLACE
LOS ANGELES, CALIFORNIA 90020

PRINTED IN USA
FIRST EDITION—APRIL 1976

TYPICAL NOTES FOR BRACING OF DUCTS*

1. — DETAILS SHOWN PROVIDE A LATERAL BRACING SYSTEM. A TYPICAL VERTICAL SUPPORT SYSTEM MUST ALSO BE USED. HOWEVER, WHERE BRACING OCCURS, THE VERTICAL ANGLE SHOWN MAY REPLACE A TYPICAL VERTICAL SUPPORT. THIS INCLUDES A TRAPEZE VERTICAL SUPPORTING SYSTEM.

2. — BRACE ALL RECTANGULAR DUCTS 6 SQ. FT. OF AREA AND LARGER. BRACE ALL ROUND DUCTS 28" IN DIAMETER AND LARGER.

3. — TRANSVERSE BRACING TO OCCUR 30'-0" o.c. MAX. (EXCEPT FOR RECTANGULAR DUCTS 61" & LARGER (IN EITHER DIRECTION) MAY BE BRACED @ 32'-0"a.) TRANSVERSE BRACING SHALL BE INSTALLED AT EACH DUCT TURN AND AT EACH END OF A DUCT RUN. LONGITUDINAL BRACING SHALL OCCUR AT 60'-0" o.c. MAX. TRANSVERSE BRACING FOR ONE DUCT SECTION MAY ALSO ACT AS LONGITUDINAL BRACING FOR A DUCT SECTION CONNECTED PERPENDICULAR TO IT, IF THE BRACING IS INSTALLED WITHIN FOUR FEET OF THE INTERSECTION OF BOTH DUCTS & THE BRACING IS SIZED FOR THE LARGER DUCT.

4. — NO BRACING IS REQUIRED IF THE TOP OF DUCT IS SUSPENDED 12" OR LESS FROM THE SUPPORTING STRUCT'L MEMBER & ATTACHED TO TOP OF DUCT

5. — A GROUP OF DUCTS MAY BE COMBINED IN A LARGER SIZE FRAME USING THE OVERALL DIMENSIONS WITH MAX. WT. FOR SELECTION OF THE MEMBERS FROM THE SCHEDULE ON SHEET 2.

6. — WALLS (INCLUDING GYP-BOARD NON-BEARING PARTITIONS) WHICH HAVE DUCTS RUNNING THRU THE THEM MAY REPLACE A TYPICAL TRANSVERSE BRACE.

7. — WHERE IT IS PRACTICAL TO DO SO, DUCTS (& PIPES) NOT BRACED, SHALL BE INSTALLED WITH A 6" MIN. CLEARANCE TO VERTICAL CEILING HANGER WIRES.

* ESSENTIAL BUILDINGS OR LIFE SAFETY EQUIPMENT. ALL SHEET METAL FOR BRACING TO BE $F_y = 33$ ksi MIN. GA. FOR SHEET METAL FOR BRACING TO BE AS FOLLOWS

     16 GA. (0.0598 INCH)
     14 GA. (0.0747 INCH)
     12 GA. (0.1046 INCH)

HILLMAN, BIDDISON & LOEVENGUTH STRUCTURAL ENGRS	GUIDELINES FOR SEISMIC RESTRAINTS OF MECHANICAL SYSTEMS.	SMACNA	MAR. 76	1

430

GENERAL NOTES FOR BRACING PIPES *—(SEE SHT. 1)

1.—BRACE ALL PIPES WITH $2\frac{1}{2}"$ I.D. AND LARGER.
2.—DETAILS SHOWN PROVIDE A LATERAL BRACING SYSTEM. A TYPICAL VERTICAL SUPPORT SYSTEM MUST ALSO BE USED. HOWEVER, WHERE BRACE OCCURS, THE VERTICAL ANGLE SHOWN MAY REPLACE A TYPICAL VERTICAL SUPPORT.
3.—TRANSVERSE BRACINGS AT 40'-0" o.c. MAX.
4.—LONGITUDINAL BRACINGS AT 80'-0' o.c. MAX.
5.—TRANSVERSE BRACING FOR ONE PIPE SECTION MAY ALSO ACT AS LONGITUDINAL BRACING FOR THE PIPE SECTION CONNECTED PERPENDICULAR TO IT, IF THE BRACING IS IN- STALLED WITHIN 24" OF THE ELBOW OR TEE AND SIMILAR SIZE.
6.—DO NOT USE BRANCH LINES TO BRACE MAIN LINES.
7.—PROVIDE FLEXIBILITY IN JOINTS WHERE PIPES PASS THROUGH BUILDING SEISMIC OR EXPANSION JOINTS, OR WHERE RIGIDLY SUPPORTED PIPES CONNECT TO EQUIPMENT WITH VIBRATION ISOLATORS.
8.—AT VERTICAL PIPE RISERS, WHEREVER POSSIBLE, SUPPORT THE WEIGHT OF THE RISER AT A POINT OR POINTS ABOVE THE CENTER OF GRAVITY OF THE RISER. PROVIDE LAT- ERAL GUIDES AT THE TOP AND BOTTOM OF THE RISER, AND AT INTERMEDIATE POINTS NOT TO EXCEED 30'-0' ON CENTER.
9.—PROVIDE LARGE ENOUGH PIPE SLEEVES THROUGH WALLS OR FLOORS TO ALLOW FOR ANTICIPATED DIFFERENTIAL MOVE- MENTS.
10.—DO NOT FASTEN ONE RIGID PIPING SYSTEM TO TWO DIS- SIMILAR PARTS OF A BUILDING THAT MAY RESPOND IN A DIFFERENT MODE DURING AN EARTHQUAKE; FOR EXAMPLE, A WALL AND A ROOF.
11.—BRACING DETAILS, SCHEDULE AND NOTES ARE TO BE USED WITH THE FOLLOWING TYPES OF PIPE: STEEL PIPE SCHEDULE 40 AND 80, COPPER PIPE TYPE K, L, M.(ONLY SILVER SOLDERED BRAZED JOINTS TO BE USED WITH COPPER PIPE)
12.—FOR GAS PIPING, THE BRACING DETAILS, SCHEDULES AND NOTES MAY ALSO BE USED EXCEPT THAT TRANSVERSE BRACING SHALL BE AT 20'-0' o.c., MAX. AND LONGITUDINAL BRACING AT 40'-0" o.c. MAX. ALSO $1"\phi$ $1\frac{1}{4}"\phi$ $1\frac{1}{2}"\phi$ AND $2"\phi$ PIPES SHALL BE BRACED THE SAME AS $2\frac{1}{2}"\phi$ PIPE IN THE SCHEDULE.
13.—ACID WASTE PIPING SYSTEMS SUCH AS GLASS, PLASTIC AND DURIRON PIPE AND CAST IRON PIPING SYSTEMS ARE EXCLUDED FROM THE GUIDELINES.
14—IN BOILER & MECH'L EQUIPMENT ROOMS, BRACING OF ALL TYPES OF PIPE (EX. STEEL, COPPER, ETC.) INCLUDED IN SCHEDULE SHALL BE EX- TENDED TO INCLUDE $1\frac{1}{4}"\phi$, $1\frac{1}{2}"\phi$ & $2"\phi$, BRACED AS REQ'D FOR $2\frac{1}{2}"\phi$ PIPE.

| HILLMAN, BIDDISON & LOEVENGUTH. STRUCTURAL ENGRS. | GUIDELINES FOR SEISMIC RESTRAINTS OF MECHANICAL SYSTEMS | SMACNA | 7 MAR.76 |

431

ADDENDUM No. 1

ADD THE FOLLOWING NOTE TO SHEET 7 :

    15.- NO BRACING IS REQUIRED IF THE TOP OF
        PIPE IS SUSPENDED 12" OR LESS FROM
        THE SUPPORTING STRUCTURAL MEMBER.

APPROVED BY :

OFFICE OF THE STATE ARCHITECT
STRUCTURAL SAFETY SECTION

APPROVED MAY 28 '76

Per *D. K. Jephcott* /s

APPLICATION NO. _____ PRINCIPAL STRUCTURAL ENGINEER

HILLMAN, BIDDISON & LOEVENGUTH

STRUCTURAL ENGINEERS *JC Loevenguth* S808

HILLMAN, BIDDISON & LOEVENGUTH STRUCTURAL ENGRS	GUIDELINES FOR SEISMIC RESTRAINTS OF MECHANICAL SYSTEMS	A1
		MAY 76

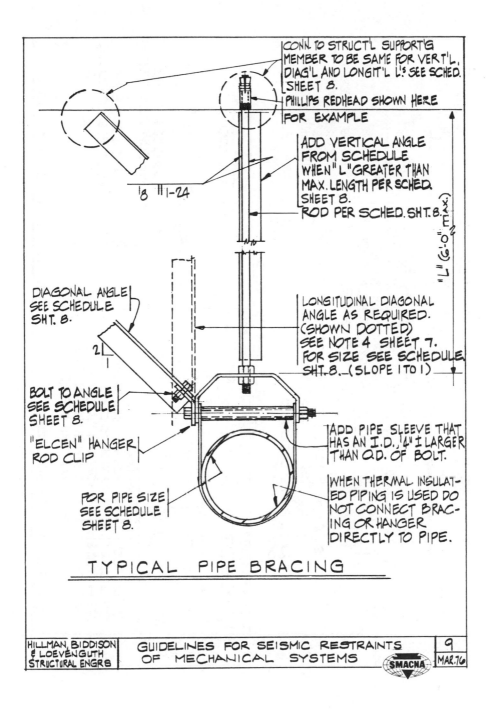

CONN. TO STRUCT'L SUPPORT'G MEMBER TO BE SAME FOR VERT'L, DIAG'L AND LONGIT'L L'S SEE SCHED. SHEET 8.

PHILLIPS REDHEAD SHOWN HERE FOR EXAMPLE

ADD VERTICAL ANGLE FROM SCHEDULE WHEN "L" GREATER THAN MAX. LENGTH PER SCHED. SHEET 8.

ROD PER SCHED. SHT. 8.

'8 #1-24

"L" (6'-0" MAX.)

DIAGONAL ANGLE SEE SCHEDULE SHT. 8.

2∟

LONGITUDINAL DIAGONAL ANGLE AS REQUIRED. (SHOWN DOTTED) SEE NOTE 4 SHEET 7. FOR SIZE SEE SCHEDULE SHT. 8. (SLOPE 1 TO 1)

BOLT TO ANGLE SEE SCHEDULE SHEET 8.

"ELCEN" HANGER ROD CLIP

ADD PIPE SLEEVE THAT HAS AN I.D. ¼" I LARGER THAN O.D. OF BOLT.

FOR PIPE SIZE SEE SCHEDULE SHEET 8.

WHEN THERMAL INSULAT-ED PIPING IS USED DO NOT CONNECT BRAC-ING OR HANGER DIRECTLY TO PIPE.

TYPICAL PIPE BRACING

| HILLMAN, BIDDISON & LOEVENGUTH STRUCTURAL ENGRS | GUIDELINES FOR SEISMIC RESTRAINTS OF MECHANICAL SYSTEMS   SMACNA | 9 MAR. 76 |

433

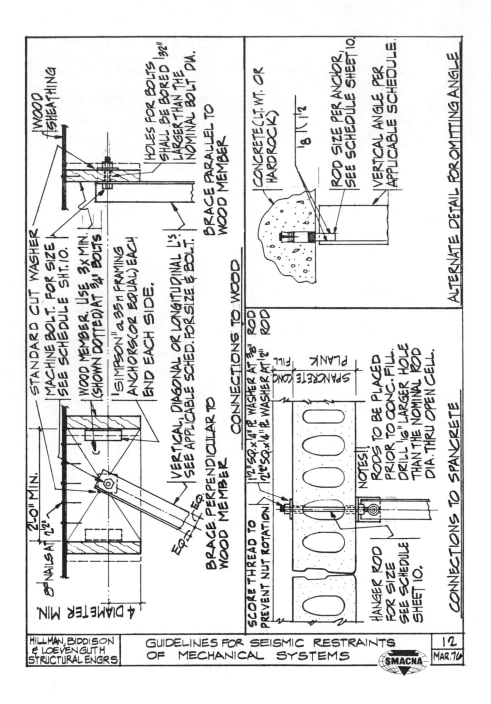

CONNECTIONS TO WOOD

STANDARD CUT WASHER

MACHINE BOLT. FOR SIZE, SEE SCHEDULE SHT. IO.

WOOD MEMBER. USE 3x MIN. (SHOWN DOTTED) AT ¾" BOLTS

WOOD SHEATHING

HOLES FOR BOLTS SHALL BE BORED 1/32" LARGER THAN THE NOMINAL BOLT DIA.

BRACE PARALLEL TO WOOD MEMBER

"SIMPSON" a 35n FRAMING ANCHORS(OR EQUAL) EACH END EACH SIDE.

VERTICAL, DIAGONAL OR LONGITUDINAL L's SEE APPLICABLE SCHED. FOR SIZE & BOLT.

BRACE PERPENDICULAR TO WOOD MEMBER

4 DIAMETER MIN.

2'-0" MIN.

1½"

8 NAILS AT 1½"

ALTERNATE DETAIL FOR OMITTING ANGLE

CONCRETE (LT. WT. OR HARDROCK)

18 1½ 1½

ROD SIZE PER ANCHOR, SEE SCHEDULE SHEET IO.

VERTICAL ANGLE PER APPLICABLE SCHEDULE.

CONNECTIONS TO SPANCRETE

1½"SQ.x4"R. WASHER AT ⅜" ROD
2½"SQ.x4"R. WASHER AT ½" ROD

SCORE THREAD TO PREVENT NUT ROTATION

SPANCRETE CONC. FILL

PLANK

NOTES:
RODS TO BE PLACED PRIOR TO CONC. FILL. DRILL 1/16" LARGER HOLE THAN THE NOMINAL ROD DIA. THRU OPEN CELL.

HANGER ROD FOR SIZE SEE SCHEDULE SHEET IO.

HILLMAN, BIDDISON & LOEVENGUTH STRUCTURAL ENGRS

GUIDELINES FOR SEISMIC RESTRAINTS OF MECHANICAL SYSTEMS

SMACNA

12

MAR. 76

434

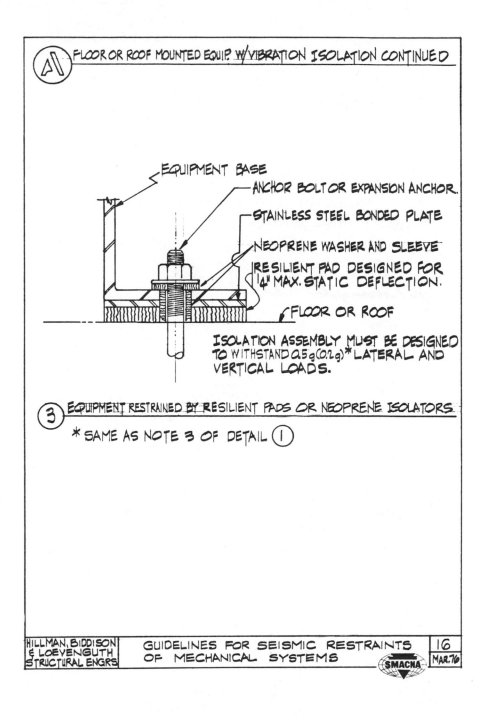

Ⓐ FLOOR OR ROOF MOUNTED EQUIP. W/VIBRATION ISOLATION CONTINUED

EQUIPMENT BASE

ANCHOR BOLT OR EXPANSION ANCHOR.

STAINLESS STEEL BONDED PLATE

NEOPRENE WASHER AND SLEEVE

RESILIENT PAD DESIGNED FOR ¼" MAX. STATIC DEFLECTION.

FLOOR OR ROOF

ISOLATION ASSEMBLY MUST BE DESIGNED TO WITHSTAND 0.5g (0.2g)* LATERAL AND VERTICAL LOADS.

③ EQUIPMENT RESTRAINED BY RESILIENT PADS OR NEOPRENE ISOLATORS.

* SAME AS NOTE 3 OF DETAIL ①

| HILLMAN, BIDDISON & LOEVENGUTH STRUCTURAL ENGRS | GUIDELINES FOR SEISMIC RESTRAINTS OF MECHANICAL SYSTEMS | 16 |
| | SMACNA | MAR.76 |

GUIDELINES FOR SEISMIC RESTRAINTS

OF KITCHEN EQUIPMENT

PREPARED FOR:

SHEET METAL INDUSTRY FUND

OF LOS ANGELES

PREPARED BY:

HILLMAN, BIDDISON & LOEVENGUTH

STRUCTURAL ENGINEERS

Jr Loevenguth SE 808

APPROVED BY:

OFFICE OF THE STATE ARCHITECT
STRUCTURAL SAFETY SECTION

APPROVED JUL 27 '77

Per *D. K. Jephcott*

PRINCIPAL STRUCTURAL ENGINEER

DATE APPROVED AUGUST 1977

REVISIONS AND/OR CHANGES TO ANY OF THE SHEETS I thru 22
WILL REQUIRE NEW APPROVAL BY O.S.A.

GUIDELINES FOR SEISMIC RESTRAINTS

OF

KITCHEN EQUIPMENT

©   S MI F

SHEET METAL INDUSTRY FUND OF LOS ANGELES
401 SHATTO PLACE
LOS ANGELES, CALIFORNIA 90020

PRINTED IN USA
FIRST EDITION – AUGUST 1977

PREFACE

These guidelines have been prepared for use by engineers, architects and contractors, approving authorities and others as an aid in standardizing details of construction for seismic restraints of kitchen equipment complying with the 1976 UBC, as modified by Title 17, C.A.C. It is not to be construed to be a design manual.

OSA approval is for anchorage only. Construction of equipment must be strong enough to resist design forces.

These guidelines were developed using sound engineering principles and judgement. They represent realistic and safe details compatible with the general guidelines and force factors in the 1976 UBC as modified by Title 17, C.A.C. They are subject to revision as further experience and investigation may show is necessary. The Sheet Metal Industry Fund of Los Angeles assumes no responsibility and accepts no liability for the application of the principles or techniques contained in this guideline.

GENERAL NOTES

1. Kitchen equipment is manufactured in a multitude of different shapes,
    sizes and weights. In order to codify the variables, equipment has
    been arranged into 19 basic categories or types; all equipment within
    a category has similar restraint requirements. Details of anchorages
    and bracing for each of the 19 basic types have been developed. The
    equipment shown in the index is not necessarily a complete list, but
    the listed equipment details may be used as a guide for similar equip-
    ment within the restraints indicated. The project architect/engineer
    should be consulted if there is any doubt about restraint requirements,
    so that allowances for bracing systems may be included in the con-
    tract documents/or contractor's bid.

2. All details in this manual are basically for equipment that is "Hardwired".
    (This means equipment connected with gas, steam, water, electrical
    lines, etc.). Plug in types are usually excluded. The exceptions are
    free-standing cabinets greater than five feet in height. The details
    in the manual show an entire assembly.

3. The details have been prepared on the basis of new construction. The
    same details are applicable to remodeling, provided the specified struc-
    tural capacities of the existing structures, such as stud walls, floors,
    etc., are equivalent or substantiated.

4. Where manufacturers' names and catalog numbers have been used as
    a reference for structural integrity, equivalent products of other man-
    ufacturers may be substituted provided documentation is supplied,
    showing the alternates are structuraly equivalent.

5. The perimeter stud walls (where equipment is to be anchored) around
    the kitchen area are 18 ga. studs with $1\frac{1}{4}$" flange width. The architect/
    structural engineer must indicate this requirement on the construction
    documents. The studs shall be securely anchored top and bottom to
    develop an equivalent horizontal force of 10 psf. Steel shall be of
    commercial quality and yield point = 33000 psi. Stud design shall be
    verified for each particular installation.

6. Finishes indicated on the details are not included in the kitchen equipment contract.

7. All embedded items and wall plates to be provided by the kitchen equipment contractor. Kitchen equipment contractor shall provide location drawings for all embeded items and wall plates. The general contractor shall coordinate the work of other trades as they relate to the installation of kitchen equipment.

8. Where backing plates occur on stud walls and dry wall is used, plastering contractor to install so as to provide a smooth surface.

9. Where concrete and/or masonry walls occur, any equipment that requires anchorage may be fastened directly to the wall with 3/8" expansion anchors in lieu of the backing plate and sheet metal screws shown on the details.

10. Expansion anchors and expansion stud anchors called for on the details are as follows:

Expansion Anchor	Expansion Stud Anchor
Hilti Drop-in *	Hilti Kwik-Bolt **
Phillips Non-Drill Anchors	Phillips Wedge Anchor *
	Phillips Stud Anchor

* (Hard Rock concrete only)
** (Lightweight concrete only)

10. Where expansion anchors are loaded in pull out, 50% of the bolts must be proof tested to twice 80% of I.C.B.O. allowable loads. If any failures occur, the anchor must be replaced and the immediately adjacent anchors must also be tested.

11. Minimum concrete strengths to develop expansion anchors.

   $f'c = 2000$ psi @ 28 days (Hard Rock)
   $f'c = 2500$ psi @ 28 days (Light weight 110 pcf)

12. All screws shall penetrate $\frac{1}{4}$" minimum through metal framing.

TYPE & DETAIL INDEX

441

442

443

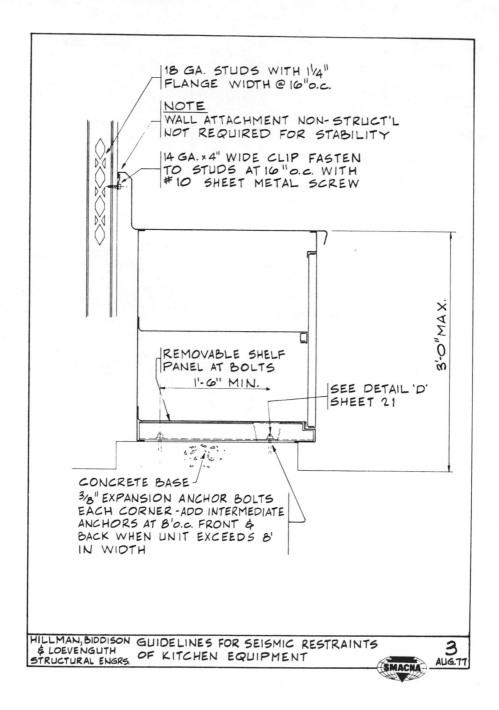

18 GA. STUDS WITH 1¼"
FLANGE WIDTH @ 16" o.c.

NOTE
WALL ATTACHMENT NON-STRUCT'L
NOT REQUIRED FOR STABILITY

14 GA. x 4" WIDE CLIP FASTEN
TO STUDS AT 16" o.c. WITH
#10 SHEET METAL SCREW

3'-0" MAX.

REMOVABLE SHELF
PANEL AT BOLTS
1'-6" MIN.

SEE DETAIL 'D'
SHEET 21

CONCRETE BASE
⅜" EXPANSION ANCHOR BOLTS
EACH CORNER - ADD INTERMEDIATE
ANCHORS AT 8' o.c. FRONT &
BACK WHEN UNIT EXCEEDS 8'
IN WIDTH

HILLMAN, BIDDISON
& LOEVENGUTH
STRUCTURAL ENGRS.

GUIDELINES FOR SEISMIC RESTRAINTS
OF KITCHEN EQUIPMENT

SMACNA

3
AUG. 77

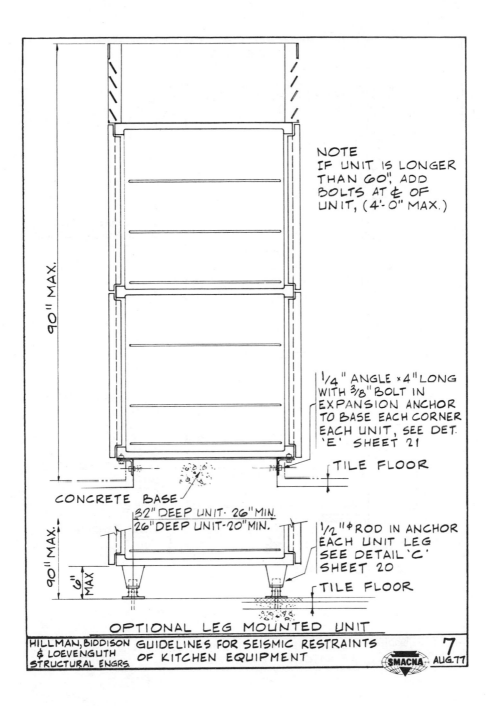

NOTE
IF UNIT IS LONGER
THAN 60", ADD
BOLTS AT ℄ OF
UNIT, (4'-0" MAX.)

90" MAX.

¼" ANGLE × 4" LONG
WITH ⅜" BOLT IN
EXPANSION ANCHOR
TO BASE EACH CORNER
EACH UNIT, SEE DET.
'E' SHEET 21

TILE FLOOR

CONCRETE BASE

32" DEEP UNIT· 26" MIN.
26" DEEP UNIT· 20" MIN.

90" MAX.

6" MAX

½"⌀ ROD IN ANCHOR
EACH UNIT LEG
SEE DETAIL 'C'
SHEET 20

TILE FLOOR

OPTIONAL LEG MOUNTED UNIT

HILLMAN, BIDDISON & LOEVENGUTH STRUCTURAL ENGRS.	GUIDELINES FOR SEISMIC RESTRAINTS OF KITCHEN EQUIPMENT	7 SMACNA AUG. 77

445

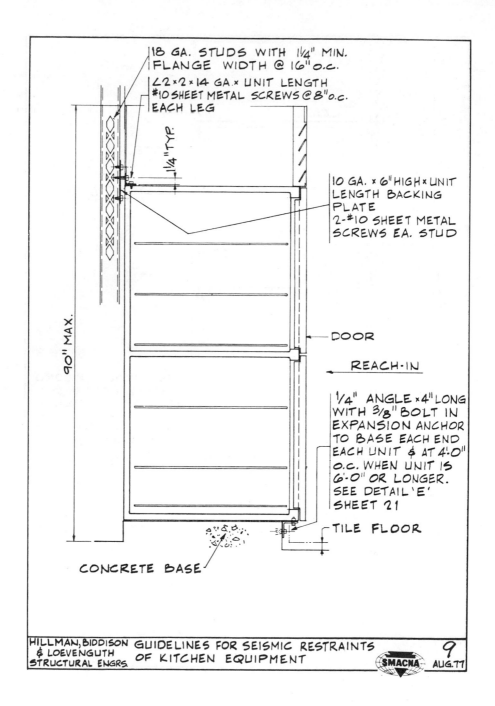

18 GA. STUDS WITH 1¼" MIN. FLANGE WIDTH @ 16" O.C.

∠2×2×14 GA.× UNIT LENGTH #10 SHEET METAL SCREWS @ 8" O.C. EACH LEG

1¼" TYP.

10 GA. × 6" HIGH × UNIT LENGTH BACKING PLATE 2-#10 SHEET METAL SCREWS EA. STUD

90" MAX.

DOOR

REACH-IN

¼" ANGLE ×4" LONG WITH ⅜" BOLT IN EXPANSION ANCHOR TO BASE EACH END EACH UNIT & AT 4'-0" O.C. WHEN UNIT IS 6'-0" OR LONGER. SEE DETAIL 'E' SHEET 21

TILE FLOOR

CONCRETE BASE

| HILLMAN, BIDDISON & LOEVENGUTH STRUCTURAL ENGRS. | GUIDELINES FOR SEISMIC RESTRAINTS OF KITCHEN EQUIPMENT | SMACNA | 9 AUG. 77 |

446

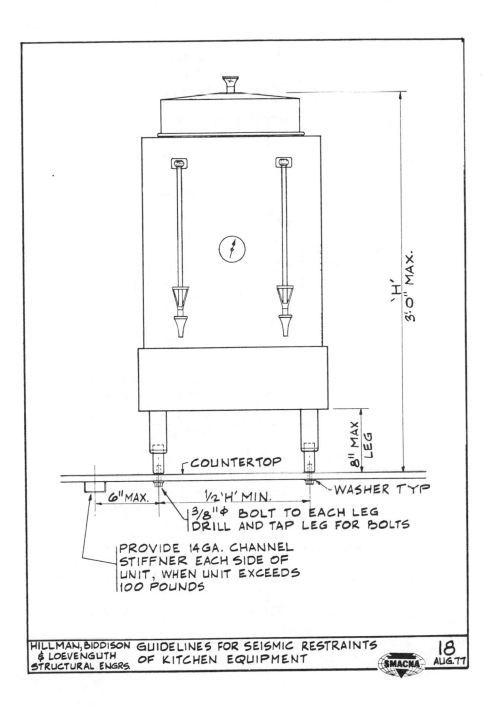

'H'
3'0" MAX.

8" MAX LEG

COUNTERTOP

WASHER TYP

6" MAX.

½ 'H' MIN.

3/8"⌀ BOLT TO EACH LEG
DRILL AND TAP LEG FOR BOLTS

PROVIDE 14GA. CHANNEL
STIFFNER EACH SIDE OF
UNIT, WHEN UNIT EXCEEDS
100 POUNDS

HILLMAN, BIDDISON
& LOEVENGUTH
STRUCTURAL ENGRS.

GUIDELINES FOR SEISMIC RESTRAINTS
OF KITCHEN EQUIPMENT

SMACNA

18
AUG.77

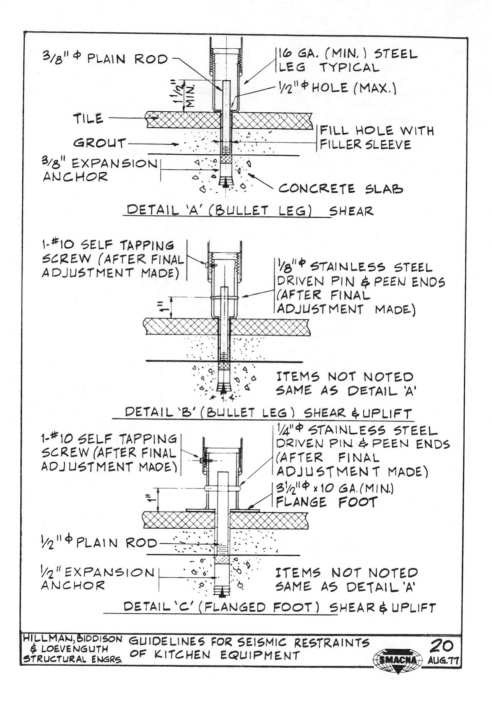

3/8"⌀ PLAIN ROD

16 GA. (MIN.) STEEL LEG TYPICAL

1/2"⌀ HOLE (MAX.)

1 1/2" MIN.

TILE

GROUT

FILL HOLE WITH FILLER SLEEVE

3/8" EXPANSION ANCHOR

CONCRETE SLAB

DETAIL 'A' (BULLET LEG) SHEAR

1-#10 SELF TAPPING SCREW (AFTER FINAL ADJUSTMENT MADE)

1/8"⌀ STAINLESS STEEL DRIVEN PIN & PEEN ENDS (AFTER FINAL ADJUSTMENT MADE)

1"

ITEMS NOT NOTED SAME AS DETAIL 'A'

DETAIL 'B' (BULLET LEG) SHEAR & UPLIFT

1-#10 SELF TAPPING SCREW (AFTER FINAL ADJUSTMENT MADE)

1/4"⌀ STAINLESS STEEL DRIVEN PIN & PEEN ENDS (AFTER FINAL ADJUSTMENT MADE)

3 1/2"⌀ x 10 GA.(MIN.) FLANGE FOOT

1"

1/2"⌀ PLAIN ROD

1/2" EXPANSION ANCHOR

ITEMS NOT NOTED SAME AS DETAIL 'A'

DETAIL 'C' (FLANGED FOOT) SHEAR & UPLIFT

| HILLMAN, BIDDISON & LOEVENGUTH STRUCTURAL ENGRS. | GUIDELINES FOR SEISMIC RESTRAINTS OF KITCHEN EQUIPMENT | SMACNA | 20 AUG.77 |

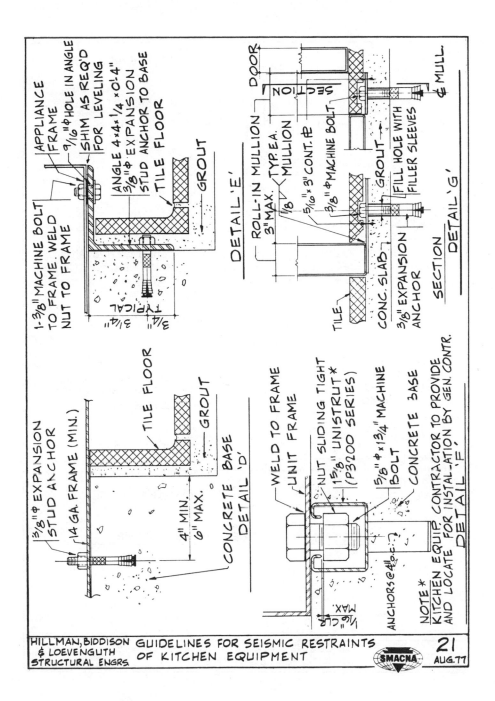

APPLIANCE FRAME

1-3/8" MACHINE BOLT TO FRAME. WELD NUT TO FRAME

9/16"Ø HOLE IN ANGLE

SHIM AS REQ'D FOR LEVELING

ANGLE 4×4×1/4×0'-4"

3/8"Ø EXPANSION STUD ANCHOR TO BASE

TILE FLOOR

GROUT

3 1/4"

3/4"

TYPICAL

DETAIL 'E'

ROLL-IN MULLION

DOOR

3" MAX.

1/8

TYP. EA. MULLION

SECTION

5/16" × 3" CONT. ℄

3/8"Ø MACHINE BOLT

CONC. SLAB

GROUT

FILL HOLE WITH FILLER SLEEVES

TILE

℄ MULL.

3/8" EXPANSION ANCHOR

SECTION

DETAIL 'G'

3/8"Ø EXPANSION STUD ANCHOR

14 GA. FRAME (MIN.)

TILE FLOOR

GROUT

4" MIN. 6" MAX.

CONCRETE BASE

DETAIL 'D'

WELD TO FRAME

UNIT FRAME

NUT SLIDING TIGHT

1 5/8" UNISTRUT* (P3200 SERIES)

5/8"Ø×1 3/4" MACHINE BOLT

CONCRETE BASE

ANCHORS @ 4'-0" O.C.

1/16" CL'R. MAX.

NOTE*
KITCHEN EQUIP. CONTRACTOR TO PROVIDE AND LOCATE FOR INSTALLATION BY GEN. CONTR.

DETAIL 'F'

HILLMAN, BIDDISON & LOEVENGUTH STRUCTURAL ENGRS.

GUIDELINES FOR SEISMIC RESTRAINTS OF KITCHEN EQUIPMENT

SMACNA

21

AUG.77

449

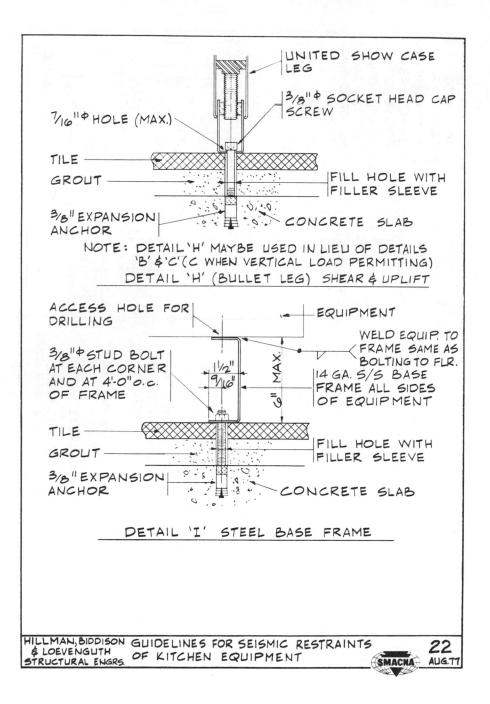

UNITED SHOW CASE LEG

3/8"⌀ SOCKET HEAD CAP SCREW

7/16"⌀ HOLE (MAX.)

TILE

GROUT

FILL HOLE WITH FILLER SLEEVE

3/8" EXPANSION ANCHOR

CONCRETE SLAB

NOTE: DETAIL 'H' MAYBE USED IN LIEU OF DETAILS 'B' & 'C' (C WHEN VERTICAL LOAD PERMITTING)

DETAIL 'H' (BULLET LEG) SHEAR & UPLIFT

ACCESS HOLE FOR DRILLING

EQUIPMENT

WELD EQUIP. TO FRAME SAME AS BOLTING TO FLR.

3/8"⌀ STUD BOLT AT EACH CORNER AND AT 4'-0" o.c. OF FRAME

1½"

9/16"

6" MAX.

14 GA. S/S BASE FRAME ALL SIDES OF EQUIPMENT

TILE

GROUT

FILL HOLE WITH FILLER SLEEVE

3/8" EXPANSION ANCHOR

CONCRETE SLAB

DETAIL 'I' STEEL BASE FRAME

HILLMAN, BIDDISON & LOEVENGUTH STRUCTURAL ENGRS.

GUIDELINES FOR SEISMIC RESTRAINTS OF KITCHEN EQUIPMENT

SMACNA

22 AUG. 77

Index

451